Project Management

Ofer Zwikael · John R. Smyrk

Project Management

A Benefit Realisation Approach

Springer

Ofer Zwikael
Research School of Management
Australian National University
Canberra, ACT, Australia

John R. Smyrk
School of Business, UNSW
Canberra, ACT, Australia

ISBN 978-3-030-03173-2 ISBN 978-3-030-03174-9 (eBook)
https://doi.org/10.1007/978-3-030-03174-9

Library of Congress Control Number: 2019932831

This Springer imprint is published by the registered company Springer Nature Switzerland AG
The registered company address is: Gewerbestrasse 11, 6330 Cham, Switzerland

To my beloved Maya, Tal and Noa

Ofer Zwikael

To Michael, Martin and Carolyn and in loving memory of Ruth

John R. Smyrk

Foreword

Finally, a book that is up to date in terms of the reality of one major project weakness that is plaguing top executives of corporations and administrators of government agencies and NGOs. *Project Management: A Benefit Realisation Approach* takes on the challenge of describing this current weakness and showing how to overcome it. The book is written by a pair of highly decorated co-authors—Ofer Zwikael and John R. Smyrk—who have won multiple awards for their groundbreaking, real-world research about the current problems facing project management today, presented in a clear, engaging, practice-oriented style.

Projects are meant to reshape operational processes of the organisation or its customers to gain some advantage. These days project management has the tools to successfully deliver the products and services desired by top management at our organisations. But the expected improvement in productivity, morale, customer delight, competitiveness or other such benefits rarely seem to be obtained, much to the disappointment of these senior executives, not to mention the lost time and expense that went into the project.

Project Management: A Benefit Realisation Approach offers a framework that starts from project selection and definition, through execution and beyond implementation until the desired benefits are realised. However, the emphasis and guidance of this book are concentrated on this last stage where project management these days is largely failing: obtaining the benefits that the project was intended to deliver. This book is a "must-read" for those managers involved at any level of their organisation with projects, particularly if the results of those projects are often disappointing. To facilitate readers' learning, the book includes both conceptual principles as well as practical tools, templates, illustrative examples, case studies and a glossary. I recommend it highly!

Winston-Salem, NC, USA

Jack Meredith
Professor of Management
and Broyhill Distinguished Scholar
and Chair in Operations, Emeritus
School of Business
Wake Forest University

Preface

While the main driver of investment in projects is the desire to realise a return in the form of a future flow of benefits, projects too often end without having achieved such a goal. This book presents an overall theory of projects, as well as discussing its practical implications, through a set of models, tools, templates and case studies.

The book has two parts. Part I proposes a novel theory of projects and a comprehensive framework for its management. Part II provides practical tools (with illustrations) for the implementation of the framework across the various phases of a project's life. It is directed at four sorts of audience: senior executives who make decisions about investments in projects, project practitioners who aim to lead projects towards a successful completion, scholars who seek to enhance theories that derive projects and students who seek reliable frameworks that can provide them an advancement in the workplace.

The authors wish to acknowledge the contributions of Elham Merikhi and thank their respective families for their patience, perseverance, encouragement and support over the past few years.

Canberra, Australia

Ofer Zwikael
John R. Smyrk

Contents

Part II Leading a Project

About the Authors

Dr. Ofer Zwikael, PMP is an Associate Professor in the College of Business and Economics at the Australian National University. The recipient of the International Project Management Association's Research Award, Dr. Zwikael is the author of three books and more than 200 scholarly peer-reviewed papers published in leading journals. In addition, he has been awarded research awards by the *Academy of Management, British Academy of Management, Emerald* and the *Australian Institute of Project Management.* Dr. Zwikael has exercised major leadership roles such as Associate Dean, Head of School, Associate Editor and on the Executive Board of three Project Management Institute (PMI) international chapters.

John R. Smyrk spent 10 years in various industries: steel-making, infrastructure, heavy engineering, chemicals and industrial instrumentation. While in private practice as a specialist in project management, he has consulted to the public and private sectors. Clients covered include manufacturing, finance, transport and government. He is currently a Visiting Fellow in the School of Business at the University of NSW in Canberra. With Ofer Zwikael, he participates in an ongoing research programme directed at the assembly of comprehensive, reliable and rigorous theoretical foundations for the discipline of project management. He was a graduate of Monash University—holding an Honours degree in Economics (with a specialisation in Econometrics) and Masters in Economics (with a specialisation in Operations Research).

Part I
Projects: A Conceptual Framework

Chapter 1
What Roles Do Projects Serve in Business?

Abstract It is through projects that organisations bring about the changes that enable them to achieve their strategic objectives. Despite this, accepted project management practice is still primarily concerned with an efficient delivery of outputs (on time, on cost and according to specification) rather than with the realisation of beneficial outcomes. We define a project as a "non-repeated planned work intended to enhance organisational performance". In this chapter, we explore the nature of projects and examine many of the current issues surrounding the way they are managed.

1.1 The Nature of Projects

Regardless of the arena in which a business operates (private, government, not-for-profit or community), its executives are under continual pressure to improve performance by undertaking projects. Whereas many operational processes are executed repeatedly, each project is undertaken only once and so we define a project as "**non-repeated planned work intended to enhance organisational performance**".

1.1.1 The Strategic Triggers for Projects

Organisations set goals and develop strategic plans to enhance their long-term performance. Projects give effect to strategic plans. Although it is necessary to align projects with organisational strategy, not all projects that an organisation funds arise directly from the demands of a strategic vision. Critical initiatives can also emerge spontaneously or opportunistically. In a similar way, Mintzberg (1994) distinguishes between *deliberate* and *emergent* strategy. There is yet a third context for new initiatives—in which they are imposed from outside the organisation. Projects can, therefore, be usefully classified by their strategic contexts:

O. Zwikael and J. R. Smyrk, *Project Management*,
https://doi.org/10.1007/978-3-030-03174-9_1

1. **Deliberate strategy implementation projects**: which give effect to an organisation's declared business strategy (such as a decision to move out of manufacturing into services).
2. **Emergent strategy projects**: arising from (often unforeseen) opportunities to enhance performance (such as acquiring a competitor who is unexpectedly experiencing financial difficulties). As well as giving effect to emergent strategy, such projects also play an important role in shaping it.
3. **Imposed projects**: those that are demanded by the environment in which the organisation operates (such as a new law requiring annual safety audits).

1.1.2 Implementing Strategy Through Projects

Regardless of the trigger, each project is undertaken by the funding organisation to generate specific intended benefits, referred to as "target outcomes". Increased quality of service, reduced staff attrition and reduced cost of production are all examples of target outcomes. Thus, a project takes on the characteristics of an investment, where resources are purchased and deployed to a project today in the expectation that target outcomes will be realised into the future. It should be noted that the investment interpretation of a project remains useful even if the outcomes being sought are non-financial. The funder of a project might, therefore, be viewed as an investor. Just as returns drive financial investment, target outcomes drive investment in projects. Because funders approve the resources dedicated to a project, it is useful to understand the factors that they consider important. Our research, summarised in Box 1.1, found that the most important reason for investment in projects is the realisation of target outcomes.

From the literature

Box 1.1: What is Important to Project Funders?
In a study conducted by the authors, the relative importance of 16 project management factors to project funders (and their funding organisations) was analysed. Table 1.1 summarises the results of this study.

Table 1.1 The ranking of project management factors according to their importance to funders

Rank	Project management factors
1.	Achieving target outcomes
2.	Approving a business case
3.	Developing a business case
4.	Developing a list of agreed outputs (deliverables)
5.	Producing outputs (deliverables)
6.	Developing a list of agreed target outcomes
7.	Effective communications with stakeholders
8.	Monitoring and controlling the project
9.	Developing a project plan
10.	Managing project risks
11.	Assigning a person accountable for target outcome achievement
12.	Support provided by senior managers
13.	Assembling a suitable project team
14.	Updating the project plan
15.	Managing the project team
16.	Developing the project team

Table 1.1 shows that realising target outcomes is the most important factor for project funders. Additional analysis confirms that the score for this factor is statistically significantly higher than the scores of other factors, thus supporting the proposition that funders treat projects as a form of investment intended to realise target outcomes.

Because of their prominence in the current literature, the concepts of "programme" and "portfolio" bear some discussion. Sometimes projects interact with each other in ways that can impact the performance of the respective investments that they represent. When this happens, funders have an incentive to coordinate the interacting projects. A collection of coordinated projects is called a programme (More is said about project coordination in Chap. 4).

The collection of projects in which an organisation has decided to invest is called a project portfolio (See Sect. 7.4). Figure 1.1 shows diagrammatically how an organisation's projects, programmes and portfolio might relate. This allows us to use the term (when appropriate) "portfolio of projects *and programmes*".

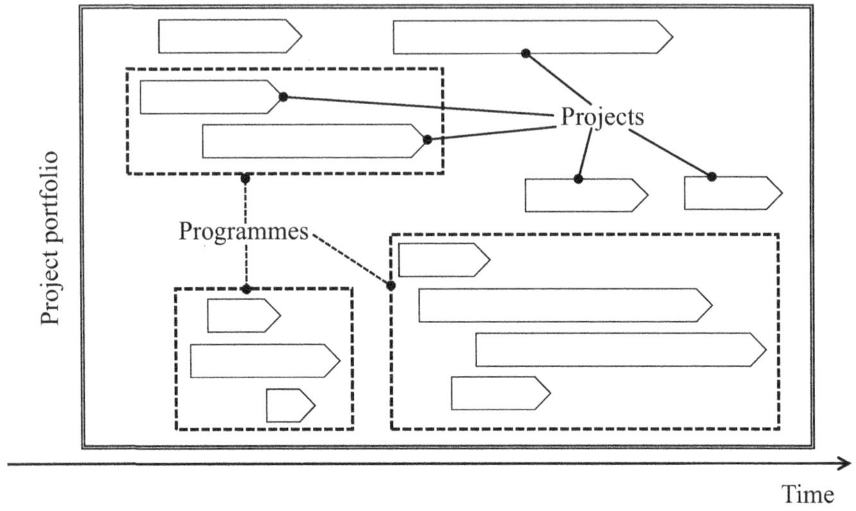

Fig. 1.1 Strategy implementation through a portfolio of projects and programmes

1.2 Trends in Today's Project Environments

The environment within which projects are undertaken has evolved over time. It is useful, therefore, to explore some of the more noteworthy trends that have helped shape the practice of project management as we see it today.

1. **Increased complexity**. Complexity can arise from a number of factors including: the size of the project, its "definability", the level of novelty, the number of stakeholders and the nature of their interest. While it is claimed that "complex" and "non-complex" projects are distinct, complexity (in all its guises) appears as a continuum. By implication, as projects become more complex, they demand increasingly intensive use of the tools, techniques and management frameworks discussed in this text.
2. **Increased globalisation**. Many projects are spread over different geographical locations and different time zones—often involving team members with a variety of nationalities. International projects, once a rare phenomenon, are now commonplace. When projects involve multiple stakeholders from different countries, the task of managing them becomes even more complex, requiring a deep understanding of cultural diversity (Zwikael et al. 2005). The phenomenon of virtual teams is one response to the issues created by global projects.
3. **The emergence of project management as a profession**. As the range of devices for planning and management has grown, so has the desire for recognition, as reflected, for example, in the establishment of professional organisations such as the International Project Management Association (IPMA) in 1965 and the Project Management Institute (PMI) in 1969. These organisations have devel-

oped project management methodologies and certification programmes for use by project managers and other key players. Examples include the Project Management Body of Knowledge (PMBOK®) (PMI 2017) and the supporting credentials, such as Project Management Professional—PMP.

Box 1.2 provides a background to some of the historical themes that have shaped today's frameworks of project management.

Reflections

Box 1.2: Historical Trends

We can infer that humans have undertaken "projects" for at least tens of thousands of years. The wonderful cave paintings at Lascaux in France certainly qualify as project outputs under the definitions adopted here. As the complexity of the work demanded by communal endeavours increased (for example, when building grand structures such as the Great Wall of China), formality became necessary for the conduct of projects—a formality that was eventually to lay the foundations of an entirely new discipline.

Whereas projects in one form or another have been undertaken for aeons, formalised approaches to their management have emerged only in the second half of the twentieth century with the term "project manager" coined by Gaddis (1959). Initially, project management focused on solving a range of particularly difficult scheduling and resourcing problems (which, with the support of the operations research community, it did very effectively). Prominent examples feature practical tools such as Gantt charts (introduced in Chap. 11).

In the latter half of the twentieth century, the business community began to accept that it too was involved with projects. Business projects are peculiar in that many of their outputs are *represented* by artefacts (rather than being artefacts in their own right). A new business process, for example, is represented by flowcharts and procedures manuals as project outputs. This class of project also brought with it a new phenomenon—in the form of ambiguous or unclear scope. The lack of a meaningful approach to this problem remains to this day as a key issue for project leadership, planning and management.

1.3 Current Challenges for Business in Project Management

The interests of those who lead projects have expanded considerably into areas such as risk management, governance, programme and portfolio management, and outcome realisation. While this expansion has enhanced our ability to manage projects, it has also presented some challenges. In what follows, we summarise the most significant of these issues and indicate where each is pursued further in the book.

1. **The project investment paradox**. Anecdotal evidence suggests that, as investments, projects have disturbingly high rates of failure. Despite this, we observe continued high levels of investment in projects. These two phenomena need to be reconciled. The performance measurement framework assembled in Chap. 8 suggests how the performance of projects should be evaluated.
2. **The emergence of programmes**. The term "programme" refers to a collection of projects. Some define a programme as collection of projects which is intended to generate target outcomes, whereby each component project is intended only to deliver outputs. This approach raises some terminological issues—for example, if a project does indeed have a target outcome, must it be relabelled as a "programme"? Chapter 4 proposes a meaningful distinction between projects and programmes and suggests a framework for the governance and management of both.
3. **The acknowledgement of project portfolios**. In the past, organisations managed simple collections of projects. Today they are being encouraged to view these collections more formally. "Portfolio" is already a well-established, clearly understood and extremely mature concept in various disciplines—particularly finance, corporate strategy and economics. The application of the concept to projects, however, has yet to mature. The term *project portfolio* is usefully confined to the collection of projects that the organisation has decided to fund. Various processes are associated with the management of a portfolio—in particular, those supporting the prioritisation and selection of projects. The relationships amongst projects, programmes and portfolios are discussed in Chap. 4.
4. **Unreliable (or non-existent) statements of project scope**. Before work on a project gets under way two questions regarding its scope must be answered by the funder: "What outputs are to be delivered?" and "What characteristics are to be built into each output?" Expressed another way, a decision has to be made before project commencement: "Of all the lists of outputs (and their required characteristics) that might be proposed, which one defines the scope of *this* project?" The existing literature assumes the list of outputs and their characteristics is "given" to the project manager at the start of the project but offers little guidance on how an appropriate supporting decision is to be taken by the project funder. This is surprising because the selection of outputs will not only determine the project's costs and timeframe, but it will also have a fundamental impact on later generation of outcomes. In Chap. 9, we propose a methodology

and tools to support the scoping process as part of the development of project business cases.

5. **Acceptance of flawed business cases**. A business case is infeasible when either of two situations exists: the project's outputs cannot be produced, delivered and implemented within the imposed constraints of time and cost, or when those outputs are insufficient for the generation of its target outcomes. Evidence suggests that impressively large numbers of projects are funded despite having infeasible business cases—which could be seen as representing perverse behaviour at best and professional negligence at worst, potentially leading to project failure. The psychology discipline suggests that decision-makers' forecasts are systematically and predictably too optimistic about the future ("planning fallacy") leading to overestimates of benefits and underestimates of cost and risk levels (Weick and Guinote 2010; Kahneman 2011). In addition to the psychological bias, some benefits are deliberately inflated ("strategic misrepresentation") to increase the chance of project funding (Jenner 2009; Flyvbjerg 2017). As a result, suboptimal and unsuitable projects are endorsed, whereas more suitable projects are rejected (Flyvbjerg et al. 2018). Chapter 9 discusses the processes that should be undertaken to confirm a project is feasible for execution before any commitment is made.
6. **Conflicting interests of stakeholders**. Every project has stakeholders who are potentially impacted by or have an impact on the exercise. In some cases, these players may have different expectations about the project. For example, certain stakeholders may want the project to address concerns that they see as important, but in which the funder has no interest. There may also be conflicting and irreconcilable views amongst stakeholders about the project goals, including some who may even oppose the project. This could very well be the case, for example, with employees whose roles will be changed or who might even face the loss of their jobs. The central issue here, however, is not one of meeting the divergent needs of stakeholders, but rather one of engaging those stakeholders whose expectations will not be met. Because project success may be significantly impacted by the behaviours, attitudes and involvement of stakeholders, frameworks of project management usually include components that seek to influence the form, direction and nature of that impact. Chapter 5 explores the concepts and techniques that underpin stakeholder management.
7. **The project manager's authority and responsibility**. The typical *function based* organisational structure often places project managers in the position where they lack the power, resources, budget and authority necessary to influence a project's results. This model does not accommodate projects well. Because however, for many organisations, project-oriented activity is still a relatively low proportion of overall business load, a *project-based* organisational structure is not appropriate. As a result, many project managers face some difficult challenges, such as power clashes with line managers, contention for the time of those team members who face dual lines of reporting and conflicts between authority and responsibility. This issue, too, can be addressed through project governance models (discussed in Chap. 4).

8. **The low levels of involvement by senior executives in projects**. Frequently a situation arises whereby the most important stakeholder in a project (the funder) has little understanding of the details surrounding the initiative—especially during execution. While one would expect that the funder knows enough from the business case and project plan to make sensible decisions about approving the exercise, a variety of factors conspire to make it difficult for him/her to gain and maintain an ongoing deep knowledge of progress. The most prominent of these factors relates to the breadth of the portfolio of investment for which the funder is responsible and the competing demands for attention from other quarters. We discuss this issue in Chap. 2 and propose (in Chap. 4) that it will be resolved through the addition of a project ownership role into the project governance model.

1.4 Issues with Current Project Management Methodologies

An outside observer of the project management discipline would find that while, over the years, it has developed many prominent characteristics, none would be as impressive as its sheer formality. This is reflected, for example, in the formation of professional organisations, the adoption of accreditation schemes, the development of recognised educational programmes and the construction of (frequently exquisite) methodologies. See, for example, the bodies of knowledge developed by the PMI, IPMA, the UK Association for Project Management (APM), and UK Office of Government Commerce (OGC) (PRINCE2). While some consider these developments as evidence for a substantial body of knowledge, others are of the view that the emerging "accepted rules of project management" may have outpaced the robustness of their theoretical underpinnings.

While much of the material that has been assembled in these (and other) methodologies has strong, proven and reliable underpinnings, they are, at the same time, surrounded by a rich and fascinating collage of accepted practice, proprietary products, agreed standards, regularised procedures, anecdotal evidence, folklore, urban myths, professional ritual, assertions, strongly held beliefs and methodological bias. Such a situation may well be explained by, what appears to be, the lack of any overarching, robust and unified "theory of projects".

Box 1.3 discusses concerns raised in the literature regarding project management theory. In what follows, we present a theoretical framework which has emerged from recent research—a framework that could go some way to addressing these concerns. This focuses on target outcome generation from the funder's perspective rather than being confined to output delivery from the project manager's point of view. Some additional issues that have triggered the need for an alternative project management framework are discussed below:

1. **The emergence of tailored project management frameworks**. Although it is claimed that frameworks of project management can be applied universally, there is a growing acceptance that "one size does not fit all". In other words, differences in project types, industries, cultures, levels of complexity and other factors influence the way in which projects should be managed. This understanding has triggered the development of subordinate project management methodologies. It is worth noting the extensions to the PMI's body of knowledge for the construction and government sectors and the development of project management approaches for specific cultures (such as the Project Management Association of Japan). The information technology sector, for example, has developed particular project management methodologies and an associated lexicon ("Agile Project Management" being a case in point). Moreover, formal project management roles, such as "planners" and "estimators", are widely accepted in the construction sector, but not recognised in other industries. All this suggests that current project management methodologies have to be adapted to different scenarios. This book proposes a rigorous project management framework that can be applied in all project contexts, while remaining flexible enough to accommodate the peculiarities of different project settings.
2. **Inconsistent and incomplete terminology**. The terms used throughout the project management profession have not been standardised. As well as different words being used to identify the one concept, particular terms are also used to identify unrelated concepts. For example, "project customers" in the service sector are usually called "end users" in the information technology sector, while "sponsor" can refer to any of a number of stakeholders such as funder, owner and champion. There is also considerable confusion with the use of "outcome" and "outputs" in various sources (Zwikael and Smyrk 2012). All the terms used throughout this text have the meanings offered in the integrated glossary provided as Appendix A.
3. **Inadequate governance models**. The central role played by target outcomes in the assessment of projects is a key thrust of the framework presented here. In general, existing project management methodologies are primarily concerned with output delivery, often neglecting outcome realisation (Zwikael and Meredith 2018). One of the significant consequences of this view is that project management governance models recognise only one accountability—related to output delivery. While it is appropriate to make the project manager accountable for the delivery of outputs that are fit-for-purpose, on time and within budget, for a number of reasons that we explore later, it is not appropriate to make him/her accountable for securing project's target outcomes. This suggests, therefore, that a new project player needs to be recognised—someone who will be held accountable for target outcomes. Chapter 2 suggests an assignment of accountability for certain results that will eventually be used as criteria when making judgements about project performance. The incorporation of this accountability in the project governance model is covered in Chap. 4.
4. **Perceptions of success**. Judgements about success/failure are commonly confined to the "iron triangle"—based on the delivery of outputs against quality, time

and cost criteria. Because this approach ignores outcomes, it would appear to be flawed and yet, at the same time, it would also appear cavalier in the extreme to abandon the conventional criteria of scope/time/cost. The "benefits realisation" models currently being proposed, do not explain satisfactorily how outcomes can be incorporated into the conventional list of success criteria. In Chap. 8, we propose a radically new approach to this issue based on three distinct judgements about project success.

5. **Myopia in the view of a "project"**. If a project is seen as concerned only with outputs, then the exercise concludes when they have all been delivered. This limited approach ignores the benefits sought from projects. Chapter 2 presents a structure in which the project ends when a decision can be made about the realisation of target outcomes.
6. **Poor understanding of the relationship between outcomes and outputs**. While the existing discussion of "benefit realisation" accepts that both outputs and outcomes emerge from projects, it sheds little light on the nature of those two concepts. Two questions have been left without satisfactory answers: What is the essential difference between an output and an outcome? How are they related? In Chap. 2, we propose a conceptual model explaining the mechanism by which outcomes are generated by a project.

From the literature

Box 1.3: Concerns about Project Management Theory

Many scholars criticise current project management methodologies and the lack of a robust theory. For example, Shenhar and Dvir (2007, p. 10) argue that "the classical drivers of project management are no longer sufficient in the current business environment". Meredith and Mantel (2015) claim that the project management literature is often characterised by non-rigorous research methods and frameworks that are unrelated to previous work. Various reasons for the relative immaturity of project management theory have been proposed. Specific criticisms include claims that the literature has been practitioner-driven and reliant on war stories (Jugdev 2004), extensive use of normative (rather than positive) approaches, and appeal to lists of factors derived from surveys of project practitioner opinions, rather than empirical research grounded in theory. In addition, a number of large and influential organisations (such as IBM, Ericsson, Motorola and Philips) have found it necessary to develop their own project management methodologies. Increasingly, there have been calls for improved theory generation through research designs that build on existing literature to develop models for rigorous, evidence-based testing in industry.

In response to this criticism, two major initiatives have been taken to make a significant and scientific contribution to existing project management methodologies. One is "Rethinking project management" in which a UK-based network group aims "to extend, enrich, reshape and develop this field beyond its current intellectual foundations" (Cicmil et al. 2006). A second case is the work by Shenhar and Dvir (2007) summarised in their book entitled "Reinventing Project Management". Yet, despite the intensity of the current debate, there is little consensus on the core principles of project management theory.

Summary

While the project management profession has assembled a formidable framework of methodologies, tools and techniques to address the challenges faced by project managers in delivering outputs, a largely unmet need exists to extend this framework to support the realisation of project benefits. In Chap. 2, we discuss a fundamental theoretical extension to the accepted frameworks of project management.

Chapter 2
A Theoretical Framework for Projects

Abstract This chapter proposes a "theory of projects" by explaining how inputs (economic resources), processes (work), outputs (artefacts) and target outcomes (closely related to benefits) are connected. The ITO (Input-Transform-Outcome) model seeks to describe the mechanisms by which outcomes are generated from outputs and, as a by-product of that discussion, explain how the two concepts are distinguished. The ITO model is also used as the foundation for various elements of the project management framework covered elsewhere in the book.

2.1 Projects as a Class of Process

Organisations operate through execution of various processes. A process is a structure of human-directed activity that produces one or more physically identifiable outputs. A new insurance policy, a new residential building, a fishing licence, a prototype of a new chemical pump, a re-engineered procurement procedure and a demolished steel plant are all examples of outputs (albeit from processes of widely varying sizes and types).

Projects are sometimes undertaken to develop a new operational process (or re-engineer an existing process). Consider, for example, an air conditioning manufacturer developing a new operational process, such as *assemble compressor*. When implemented, the new assembly process will be executed repeatedly in some manufacturing environment (using the same operational script each time). By way of contrast, the development of that new operational script will involve a project because, presumably, there is no existing description of how that development is to be carried out. It is important to note that two different scripts are associated with ventures of this kind. In this case, one will (eventually) take the form of an instruction manual which will describe in detail the tasks to be undertaken by workers on the production line as they assemble the compressor (an operational process), while the other script describes the tasks to be undertaken by the team charged with developing that instruction manual. The first script will be executed many times as each compressor is built while the second will be undertaken only once—as part of the project to develop the instruction manual.

O. Zwikael and J. R. Smyrk, *Project Management*,
https://doi.org/10.1007/978-3-030-03174-9_2

It is common to hear people argue that "Everything is a project", but, as Box 2.1 suggests, this is based on a misconception about the differences between projects and routine operations.

Practicalities

Box 2.1: A Popular Confusion: "Everything is a Project"
The popular claim that "Everything is a project" implies that all frameworks and protocols surrounding a project should be applied to all work. Moreover, Jensen et al. (2016) claim that projects are a "human condition" (something universal and stable over time), which results in many of us organising our lives by projects—from a family vacation to a career move. As a project society, the authors claim that projects are omnipresent as a form of coordinating human activities. Sometimes this is taken even further, with organisations embarking on initiatives to "projectise" all their operational processes.

A little thought reveals that, although usually well-intentioned, exercises of this kind are rather problematic. Consider the regular processing of a customer's trolley at the supermarket checkout. It would be a cause of some surprise if, on arrival at the checkout, the staff member announced that it would be necessary to obtain approval for and plan all the work involved in serving you! Clearly, unlike projects, operational processes in general have an existing script for their execution and so a project is not required. This general proposition is in no way weakened by the fact that, at some time in the past, there may well have been a project to write the script for the current checkout process.

The executions of projects and operational processes are both (at least in certain critical respects) similar—consuming resources, demanding work and producing outputs. Presumably, we could extend this list of shared features even further to include many other components such as a framework of risk management. What can we conclude from this similarity? All we can infer is that regular day-to-day business operations and project execution are both examples of *processes*. Clearly although "business-as-usual" is a special case of a process, and a project execution is a special case of a process—the conclusion (that everything is, therefore, a project) is a *non sequitur*.

Now having said all that, it does indeed make a lot of sense for organisations to formalise the frameworks of management they use for their routine processes, but in doing so, they should be more concerned with "operationalising" these processes rather than "projectising" them.

2.2 The Input-Process-Output (IPO) Model

This section formalises the Input-Process-Output (IPO) model, discusses the limitations of such view of a project and presents the foundations for the extended project framework that follows. As discussed in Chap. 1, the work involved in project execution is a special class of process; hence can be represented by an IPO model, made up of:

1. **Inputs**. The economic resources consumed by a project—broadly categorised as *labour* (of the project team) and the *money* required for purchased resources (such as raw materials).
2. **Process**. The work done during project execution to produce its outputs.
3. **Outputs**. The physical results of a project—also known generically as *deliverables*.

The IPO model provides us with a simple but extremely potent conceptual view of project execution—as shown diagrammatically in Fig. 2.1.

A trichotomy can be imposed on organisational work of: ad hoc tasks, operational processes and projects—each requiring its own peculiar framework of management. The taxonomic tree in Fig. 2.2 shows how a process can be categorised according to this trichotomy.

A walkthrough of three examples will help in understanding the trichotomy that underpins it.

- I am about to attend a meeting in a nearby building which I have never seen before. I can if I wish, decide simply to walk there and figure out where the room is "on the fly". As the entity providing the resources (e.g. my time) for the "walking" process, I am "funding" it and deem it unnecessary to develop a script before walking to the meeting—"making it up as I go along"—thus treating the (walking) process as an ad hoc task.
- I am about to prepare a casserole—for which I have a detailed and reliable recipe. Because of the consequences of getting wrong (wasted food, late meal, embarrassed guests and so on), I deem it necessary to have a script. I would, therefore,

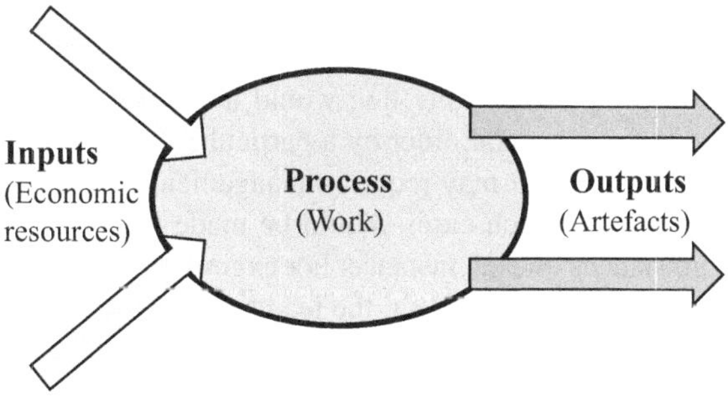

Fig. 2.1 The Input-Process-Output (IPO) model of project execution

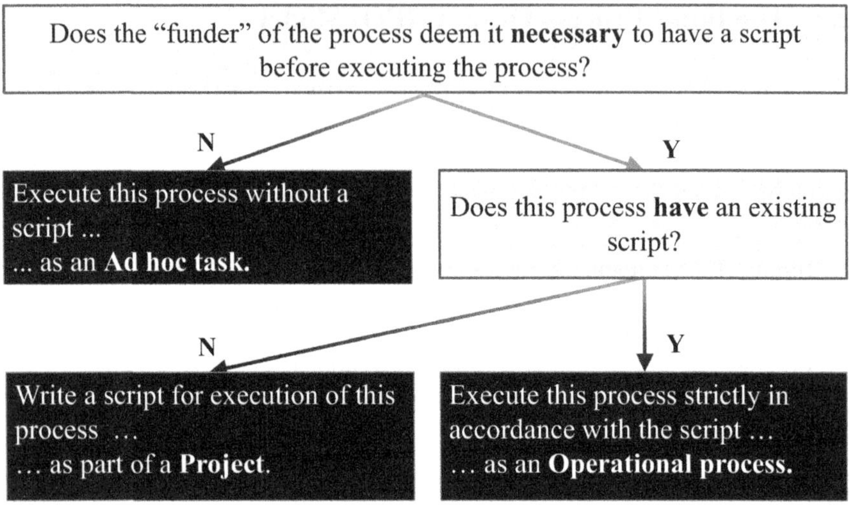

Fig. 2.2 The three types of organisational work

treat the work of making the casserole as an operational process—using the recipe as a script.

- I am about to renovate a kitchen—in accordance with a specification prepared by an interior designer. The specification tells me about colours, materials and appliances—but nothing about the work involved. Because I judge the consequences of getting it wrong as unacceptable (frustration, cost and delays) I deem it necessary to have a script. Figure 2.2 indicates that this is to be undertaken as a project. Because there is (as yet) no script, I am then faced with two (sequential) "blocks of work": the first one will involve assembling the (missing) script, while the second will involve the actual renovation of the kitchen (in accordance with the newly assembled script). The assembly of a script and its later execution are two (of four) core components of the life of a project. (In Chap. 3 we examine the way these two components fit into the overall structure of a project).

It is important to note that, in certain situations, a specific execution of an operational process may have to be reclassified using the taxonomic tree represented by Fig. 2.2. For example, manufacturing of school desks for retail distribution using an automatic assembly line would normally be treated as an operational process. However, a special order by a particular city council of 1,000 desks with a slightly different design may require management as a project. A final decision about the treatment of such cases should be made by the organisation given the unique circumstances of each instance. For example, issues may arise related to such things as the design of the product, the lead time for the order, the organisational structure of the manufacturing department, its project management maturity, the importance of the process and the standing of the purchaser. Eventually, this work will have to be executed as either an operational process or a project.

All work (including ad hoc tasks, operational processes and the execution of projects) can be represented in this way. The IPO model implies a chronology (left-to-right) where, in turn: resources are acquired and deployed to the work of the project, the work is executed to produce certain outputs and those outputs are then delivered to the outside world.

Other (distinct) diagrammatic representation of processes are also known as "input-process-output models"—such as the Data Flow Diagram (DFD) used in systems analysis. The DFD is a data manipulation process model where data elements rather than economic resources are recognised as the inputs. Such alternative representations arise because different disciplines require different analytical devices. While a DFD could be readily presented as an IPO model along the lines of Fig. 2.1, there would be little point in doing that because it is concerned with the analysis of data flows, not with the analysis of resources consumed during the execution of the data manipulation process.

Similar to the approach discussed in Fig. 2.2, Harvey and Aubry (2018) distinguish between various types of processes. They refer to operational routine processes (when no expert judgement is involved) as either simple or complicated. Further, they refer to complex processes and projects when expert judgement is involved. Although projects are more unique than complex processes, there is still a level of overlap between the two.

2.3 Project Outputs

2.3.1 Forms of Outputs

The IPO model accepts that outputs are produced by the work carried out during project execution. A useful distinction can be drawn between two types of physical result of a project:

1. **An artefact**—as would be the case in a project to build a new steel-making plant. Here the output is the new plant.
2. **A "null"** (something that no longer exists)—as would be the case in a project to demolish an old abandoned steel plant. Here the output is the "eliminated" plant.

It is noteworthy that some of the most costly projects that the world faces over the next two or three decades have null outputs. We refer here to nuclear power plants that are nearing (or have already reached) the end of their economic lives. For example, the costs of decommissioning Sellafield (a nuclear fuel reprocessing and nuclear plant decommissioning site which was built in the UK during the 1950s) are expected to exceed USD B50 (Invernizzi and Locatelli 2018).

The outputs from a project must be either an artefact or a null. Before they can be produced, it is necessary not only to *identify* all outputs from a project but also to *define* them. Outputs are defined by characteristics called fitness-for-purpose features

(FPFs). A fitness-for-purpose feature is a special case of a *requirement*. Because "requirements" in popular parlance include concepts other than fitness-for-purpose features, we do not use the term in this book. The project described in Box 2.2 illustrates how an output is defined in terms of its fitness-for-purpose features.

To produce an output, it is not only necessary that it has been identified and defined—but that it has also been *specified*. A specification is based on an output's definition, but significantly expanded so that a reliable, comprehensive script for the production of the output can be assembled before work starts. The process of *specifying* an output takes place during project execution.

Practicalities

Box 2.2: The Concept of Fitness-for-Purpose Features a Case Study: Historic Australian Gold-Mining Township

An Australian state tourism authority has decided to promote a historic gold-mining township to middle-class Chinese tourists. During the mid-1800s, Chinese flocked to the newly-discovered goldfields to work as prospectors and miners.

The township has been classified by the National Trust and so there are restrictions on what can be done. For example, signage must "harmonise" with the existing architecture.

The project aims to increase awareness of Australia as a tourist destination amongst Chinese and increase visitation to the township ("Visitation" is a defined term in tourism industry parlance meaning the overall numbers of visitors to a target location). To achieve these objectives, it was decided to develop the following three outputs: a "heritage trail", interpretive plaques at points of interest and refreshment stalls.

Even superficial analysis quickly reveals the following as candidate fitness-for-purpose features for each of these outputs:

A. **Heritage trail**:

- Signs in both English and Chinese.
- To include points of historic interest (such as old graveyards).
- To accord with Feng Shui beliefs.
- To meet National Trust conditions.

B. **Interpretive plaques at points of interest**:

- Text in English and Chinese.
- Roofed.
- Non-slip pavers.
- Clear plastic protective overlay.
- Clearly visible from heritage trail.

C. **Refreshment stalls:**

- Good quality Australian-type merchandise and light refreshments.
- Undercover seating for 12 people.
- Photos of old workings featuring Chinese miners.

The assembly of an appropriate list of fitness-for-purpose features for an output is not a trivial matter—requiring formal analytical techniques (which are discussed in Chap. 9).

2.3.2 *The Concept of an Operand*

To develop an output, project execution operates on certain physical objects which are called *operands*. Operands are, in general, of three forms:

1. **An artefact**—such as an existing bridge that is being strengthened to allow use by heavy trucks.
2. **A natural object**—such as a channel cut by a river along an estuary that is being expanded for use by large ships.
3. **A "null"**—where, no identifiable artefact or object is the subject of the process, for example, empty land where a road will be built.

While loosely, an operand might be regarded as an "input" to a project, it does not meet our definition (an economic resource which is *consumed* by a project) and so needs a distinguishing label.

2.3.2.1 Types of Work to Create Outputs

Project work can be broadly classified by considering the (six) combinations of (three) forms of operand and (two) types of output. The result of the analysis in Table 2.1 suggests three types of work can be undertaken in a project:

Table 2.1 Projects categorised according to the type of work involved

	Output type	
Operand	Artefact	Null
Artefact	Modify	Eliminate
Natural object	Modify	Eliminate
Null	Create	N/A

1. **Create** a (new) artefact, for example, construct a new bridge.
2. **Modify** an (existing) artefact or a natural object, for example, strengthen an existing bridge.
3. **Eliminate** an (existing) artefact or a natural object, for example, demolish an existing bridge.

Because outputs are physical effects, they are always named with nouns, for example, a "new *bridge*", "strengthened *bridge*" or "demolished *bridge*". These are, respectively, the outputs from: create (new bridge), modify (strengthened bridge) and eliminate (demolished bridge). This naming convention is an example of a *word structure*. Word structures are simple rules by which various project management terms (such as an output) are labelled so that they are clearly distinguished from others (such as an outcome).

2.4 Project Outcomes

2.4.1 Outputs Versus Outcomes

Our exploration of the concept of an outcome begins with a discussion of *target* outcomes—which is then generalised to consider other sorts of outcome.

All projects are intended to produce two distinct sorts of *results*: a collection of committed outputs and a set of target outcomes. While the outputs to be delivered from a project are a critical result, even more important to the funder (the entity approving the allocation of resources to the exercise) are its end effects—known as outcomes. Whereas an output is always a tangible physical effect, an outcome is always an intangible (but measurable) end effect. Tangible means, "can be touched", and so tangibility is a required attribute of an output. "Measurable" means "can be measured" and is a required attribute of a target outcome. Outputs can also be "measured"—but in a subtly different way to outcomes. Because they are physical effects, outputs have observable attributes. For example, in a project to develop a dredge, the slurry pump could be "measured" in terms of discharge rate (m^3/hr) and motor speed (rpm). The measurable attributes of an output are called its fitness-for-purpose features. Because outcomes are end effects rather than physical effects, they should not be described as tangible.

Outcomes can be desirable or undesirable. Of particular interest to funders are a project's target outcomes—those desirable outcomes which represent the return on the investment in the project. Target outcomes can be expressed as a change in the value of a variable associated with the desired end effect, for example: "decreased waiting times for service", "increased market share" and "decreased operating cost". Table 2.2 summarises the differences between outputs and target outcomes.

Whereas an output cannot exist unless the work of creating it has been undertaken, the realisation of target outcomes can happen serendipitously without a project being undertaken. These differences are reflected in the terminology we use—whereby

Table 2.2 Outputs versus target outcomes from a project

Characteristic	Outputs	Target outcomes
Example	New bridge	Decreased travel time
Form	Physical effect	End effect
Tangibility	Tangible	Intangible (but measurable)
Intent	*What* is to be produced by the project?	*Why* is the project being undertaken?
Naming	(Possibly qualified) noun (e.g. a building)	Involving a participial adjective (a word ending in "-ed", i.e. "increased" and "reduced")
Creation mechanism	Outputs are "produced" or "delivered"	Target outcomes are "realised" or "generated"
Manageability	Output production can be controlled	Outcome realisation can only be influenced
Measurement	Through variables (with agreed units and dimensions) attached to fitness-for-purpose features (attributes)	Through variables (with agreed units and dimensions) attached to end effects
Emergence	Impossible without execution of a specific script	Possible even if project is not executed
Lead time	Available immediately after project is executed	Delayed until sometime after project execution

outputs are (variously) produced, delivered and implemented, while outcomes are achieved, generated or realised (thus suggesting a level of uncertainty in their emergence).

2.4.2 *Key Categories of Outcomes*

As well as *target* outcomes, a project can result in other types of end effects. In general, four categories of outcome are of interest to project funders—as shown in the following taxonomy:
Outcome: a measurable end effect attributable to a project:

1. **Desirable outcome**: if a change in the value of a measure for an outcome makes a project more attractive to a funder, then it is classified as desirable. As will be discussed in Sect. 2.4.3, desirable outcomes give rise to a flow of benefits. They are usefully further split into two sub-categories:

    a. **Target outcome**: a desirable outcome which the funder seeks to improve by a target amount through investment in the project. Target outcomes can emerge only after project execution.

b. **Fortuitous outcome**: a desirable outcome which the funder did not seek to improve through investment in the project. Fortuitous outcomes play no part in an investment decision. Fortuitous outcomes can emerge either during or after project execution. A case study involving this sub-category is presented in Box 2.3.

2. **Undesirable outcomes**: if a change in the value of a measure for an outcome make a project less attractive to a funder, then it is classified as undesirable. Undesirable outcomes can emerge either during or after project execution. As discussed in Sect. 2.4.3, undesirable outcomes give rise to a flow of disbenefits. They are usefully further split into two sub-categories:

   a. **Anticipated undesirable outcome**: an undesirable outcome which is known prior to a start on a project. Anticipated undesirable outcomes are taken into account in any investment decision. If the expected level of the measure for this outcome is acceptable, then the project will be funded. If the expected level of the measure for this outcome is unacceptable, then the project will not be funded.
   b. **Unanticipated undesirable outcome**: an undesirable outcome which is not known prior to a start on a project. Clearly, unanticipated undesirable outcomes play no part in an investment decision.

In summary, we can group outcomes to four categories based on two questions: "Is the outcome anticipated?", and "Is the outcome desirable?", as presented in Table 2.3. Note that anticipated and unanticipated undesirable outcomes are associated with disbenefits. The categories of *fortuitous* and *unanticipated undesirable* can also be regarded as examples of *emergent* outcomes, a concept related to "emergent strategy" commonly discussed in strategy. An illustrative example of fortuitous outcome is provided in Box 2.3.

Table 2.3 The outcomes taxonomy

		Does the funder consider this outcome to be desirable?	
		Yes	No
Is this outcome anticipated?	Yes	Target outcome	Anticipated undesirable outcome
	No	Fortuitous outcome	Unanticipated undesirable outcome

Illustrative example

Box 2.3: Fortuitous Outcomes: The LAF Business Process Improvement Project

A project undertaken by a large loss adjusting firm (identified here as "LAF Inc") illustrates the way that fortuitous outcomes should be handled when considering project performance. Loss adjusting is the profession involved in the assessment of claims on insurance policies. The processes of loss adjusting are labour-intensive, using highly paid specialists.

For some time, the loss adjusting profession has been coming under pressure from underwriters to reduce fees, and so profitability has been under threat. A few years ago, LAF responded by undertaking a significant project "Phoenix" to re-engineer its core business processes with two target outcomes: reduced operating costs (benefiting the firm itself) and reduced claim processing times (benefiting the underwriters, LAF's clients). A business case for "Phoenix" was accepted by LAF's CEO, and the project eventually completed (but significantly over budget and over time). Furthermore, the levels of reduction in both operating costs and claims processing time fell significantly short of their targets in the business case.

Near the end of the exercise (in an unrelated move), the major insurance companies went out to tender for their loss adjusting services. LAF's bid for this work was successful, largely because of the emergence (at that point) of fully documented, pilot-tested processes as output from project Phoenix.

The fortuitous eventual outcome "increased revenue" was, of course, not included in the business case that had been accepted by the CEO (because such an outcome was completely unknown to anyone) but was recognised in the final evaluation (ex post assessment).

Because they are unknown to the funder prior to project start, fortuitous and unanticipated undesirable outcomes play no part in the original investment decision. Target and anticipated undesirable outcomes, on the other hand, are known before the project starts and hence play a central role in that decision.

When gauging the actual return on the investment in a completed project, all classifications of outcomes are taken into account. For example, in Box 2.3, despite the fact that project Phoenix fell well short of expectations, because the CEO would have accepted a business case that foreshadowed events exactly as they unfolded—including the (fortuitous) increase in revenue—the project must be judged as a satisfactory investment.

2.4.3 *Benefits and Outcomes*

Target outcomes are revealed by considering the project's purpose—as reflected, for example, in the question: "What end effect is the funder seeking from the exercise?" We focus on the views of funders because they are the projects' investors. Consider the construction of the Burj Khalifa in Dubai, the tallest structure in the world. The rationale for undertaking this exercise can be uncovered by asking the question: "Why was the tower built?". The answer appears to cover a variety of objectives including those related to rental income, the standing of UAE as a nation and the promotion of Dubai as a tourist destination.

Typical answers to the question "why should we fund this project?" for three other illustrative projects are summarised in Table 2.4.

Because target outcomes might be described as "beneficial end effects", they are closely related to "benefits". A benefit is the "flow of value" to an entity (not necessarily the funder) *arising from achievement of a target outcome*. For example, in a project undertaken recently in Australia, a target outcome was reduced incidence of child sexual abuse. A reduced incidence of child abuse results in (unknown and unidentifiable) families being better off (that is, receiving a benefit) because they will now not suffer the consequences of this crime.

The underlying purpose of a project is to generate specific "flows of value" to identified beneficiaries. The strength of the relationship between target outcomes and its desired benefits is peculiar to each project. Returning to the child sex offenders project, the linkage is indirect—with, what could be, very significant lags between realisation of the target outcome and the flow of value to certain families. In general, direct measurement of benefits is difficult and so surrogate variables are assembled which yield more readily to analytical methods. These surrogate variables become the project's target outcomes. Clearly, it is important that a plausible causal relationship is established between the achievement of target outcomes and the desired "flows of value" to beneficiaries. In the case of the child sex offenders project, a strong causal relationship can be established between a reduction in the incidence of child abuse (outcome) and the welfare of those families who now avoid the consequences of that abuse (benefit). In more simple cases, the target outcome is also the benefit. For example, consider a project to re-engineer a company's procurement processes.

Table 2.4 End effects sought by funders from their investments in three illustrative projects

Project	Illustrative end effects sought by funders from their projects
Drawing up of a long-term contract with a major supplier	Reduced frequency of supply interruption or price uncertainty
Re-engineering an existing procurement process	Reduced purchasing costs
The development of a prototype dredge	Increased reliability of a decision to release a new product to the dredging industry

Here the target outcome (reduced costs of procurement) and the (financial) benefit flow to the company are identical.

Target outcomes bear further detailed discussion because, although they are not displayed in the IPO model, they do represent, nevertheless, the very reason for producing outputs.

2.4.4 The 2NY Map for Target Outcome Definition

To set the target outcomes from a project, a more detailed background discussion is required—from which emerges a supporting analytical device. Take, for example, a local government faced with an investment decision regarding a project to reduce traffic volumes in a city's Central Business District. As proposed, the project's outputs include a cross-city tunnel, changes in the configuration of existing city streets, a tolling system, a suite of management/maintenance processes and a new business unit to operate the facility. The decision of whether to fund the project in this case is based on the values of two environmental measures that have strategic importance for the funder: air pollution and noise. Three potential scenarios surround this project; each of which describes how the world is shaped (or at least that part of the world relevant to the project):

1. **A "Now" scenario**: describing the current position in which the government now finds itself. This relates to the present state of affairs, characterised by the actual values observed in the two environmental measures.
2. **A "No" scenario**: describing the future position in which the government would find itself if it decides not to approve the project. Like the other two, the "No" scenario would also be characterised by peculiar values for each of the two environmental variables. These values could well be different from those found in either of the other two scenarios.
3. **A "Yes" scenario**: describing a future position in which the government seeks to find itself as the result of investing in the initiative. This relates to the desired state of affairs, where the same two variables would take on targeted values (presumably lower than those associated with the "No" scenario) in response to the project being approved and completed.

In summary, each of the two environmental measures (air pollution and noise) has three scenario-based values ("Now", "No" and "Yes"). The three scenarios can be displayed in a diagram—as suggested below. We call this the Now-No-Yes scenario diagram (or *2NY Map* for short). Figure 2.3 illustrates the relationships amongst the three scenarios for the two measures in the vehicular tunnel project. AHQI (Air Health Quality Index) is a standard measure of air quality, while dB (the decibel) is a unit of sound intensity. The current levels of these measures (the "Now" scenario) are 7 AHQI for pollution and 80 dB for noise. It is expected that if the project is not undertaken, these levels will rise because of an increase in the number of cars on the road. As a result, the "No" scenario in the diagram has the predicted values of

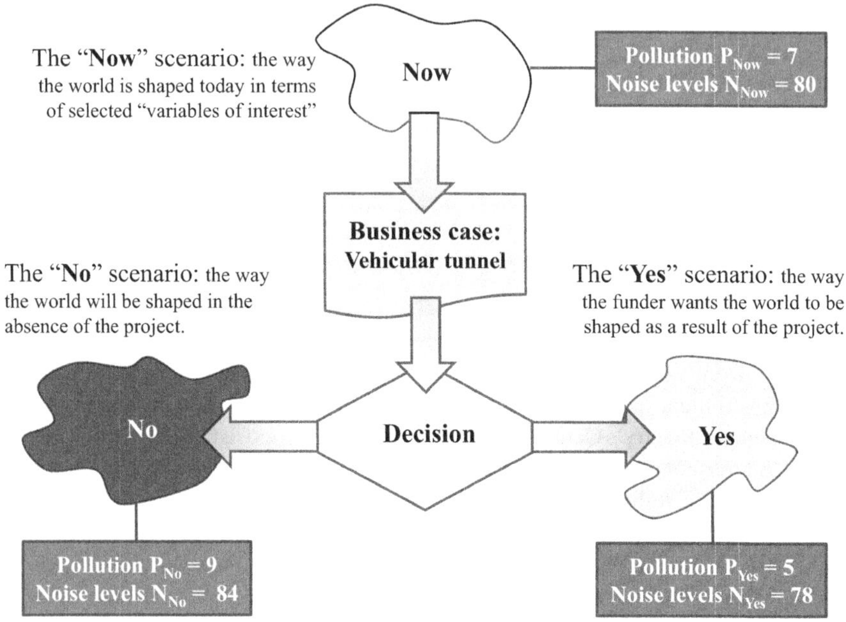

Fig. 2.3 The 2NY map (showing Now/Yes/No scenarios) for the vehicular tunnel project

9 AHQI and 84 dB, respectively. If, on the other hand, the project goes ahead (the "Yes" scenario), it is forecast that these values will go down to 5 AHQI and 78 dB, respectively.

The 2NY Map can now be used to set target outcomes for specific projects. When expressed in relative terms, these outcomes take on the differences in the values of those same variables between the "No" and "Yes" scenarios. For example, target outcomes of the vehicular tunnel project are "reduced level of pollution" (from 9 to 5 AHQI) and "reduced noise levels" (from 84 to 78 dB). Another illustration of the 2NY map (for a hospital project) is discussed in Box 2.4.

Illustrative example

Box 2.4: Setting Target Outcomes: A Hospital Process Re-engineering Project

To illustrate how the three scenarios ("Now", "Yes" and "No") are set and corresponding values established for target outcomes, consider a project to improve the public health service through the re-engineering of core hospital-related business processes. In this example, the one desired change could be *reduced average waiting time for elective surgery*. The values of this performance variable differ amongst the "Now", "Yes" and "No" scenarios in the following way:

1. The "**Now**" scenario is driven by current practices and procedures that cause the waiting time to take on a (presumably) high value. As reported by the hospital, the current average waiting time is 150 days.
2. The "**No**" scenario anticipates a world in which current practices and procedures are being overwhelmed by the load of an ageing community and so the waiting time and cost variables might be expected to grow even larger. In this case, if the project is not undertaken the average waiting time will rise to 160 days.
3. The "**Yes**" scenario is envisaged as a world with different practices and procedures that cause the waiting time and cost variables to take on a (desired) low value. It is expected that following project completion, waiting time will be 90 days.

Consistent with the discussion above, the target outcome from the proposed project—"reduced average waiting time for elective surgery" (from 160 to 90 days)—is set as the difference between the "Yes" and "No" values of this performance measure.

In light of this interpretation, target outcomes are *scenario*-related not *time*-related effects. In other words, "reduced pollution" does not mean (necessarily) that pollution levels will fall between now and when the tunnel is fully operational (a difference between "Now" and "Yes"). Instead, "reduced pollution" means that it will be less than the levels experienced had the tunnel not been built (a difference between "Yes" and "No" scenarios). Counter-intuitively, it is quite conceivable that the project is a worthy investment even though pollution levels will actually rise between now and when the tunnel is fully operational. Take, for example, a variant of the project appearing in Fig. 2.3 with the new values as shown in Fig. 2.4: In this case, at project completion, pollution and noise levels are expected to be higher than those experienced now, but lower than had the tunnel not been built. By definition, the target outcomes of the vehicular tunnel project in this case would be correctly expressed as "reduced level of pollution" (from 9 to 8 AHQI) and "reduced noise levels" (from 84 to 82 dB). The values sought for target outcomes are shown in the 2NY map as the "horizontal" differences in the project's "variables-of-interest".

For completeness, Fig. 2.4 also shows the corresponding apparent time-based change (between the "Now" and "Yes" scenarios) as the "vertical" differences in the project's variables-of-interest. The analysis confirms that, despite the apparent degradation in the variables-of-interest, the project may still represent an attractive investment. With that said, if, for some reason, time-based change is still of casual interest to some key players, care must be exercised when presenting this information. Consider, for example, a completed project where the variables-of-interest had "worsened" from before the project started. Based on that information alone it would

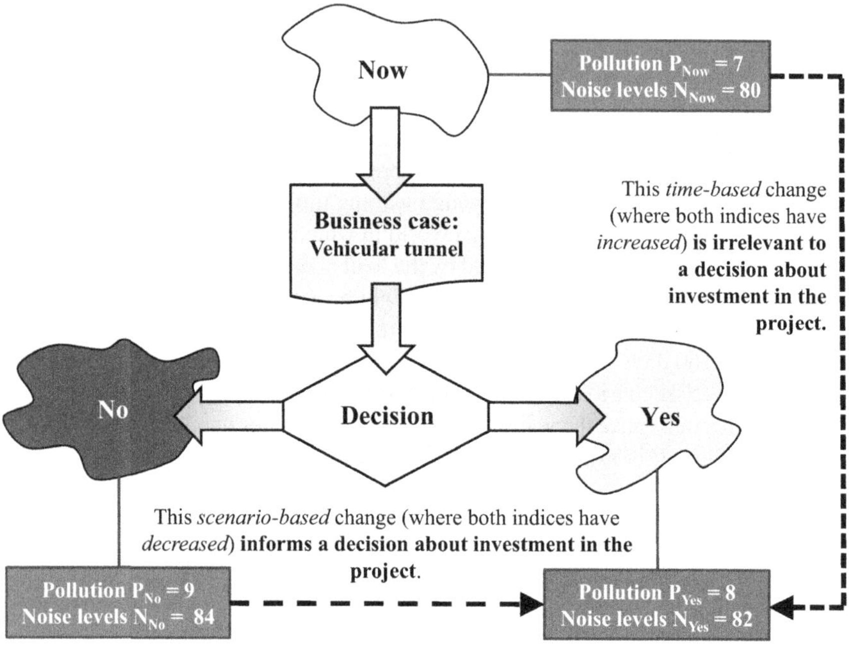

Fig. 2.4 Why time-based change is irrelevant to project investment decisions

be incorrect to claim that it should not have been funded in the first place or even that it failed. To better understand the concept of the 2NY Map and how it can be used effectively in practice, see Box 2.5 for a discussion of special cases of this tool.

Reflections

Box 2.5: Special Cases of the 2NY Map

The discussion above is concerned with the most general case in which the "Now", "Yes" and "No" scenarios are distinct. There are situations, however, whereby, the "Now" scenario can be used as a surrogate for either the "No" or the "Yes" scenario. In both cases, the "Now" values of the performance variables can then be used as estimates for their values under either of the other two scenarios. Below is a discussion of three special cases where two of the scenarios are identical in their performance variables values:

1. **Now ≡ No.** It is assumed here that, without intervention, the "Now" scenario will remain as the (undesirable) "No" scenario. In this case, the values of the performance variables will not change between the "Now" and "No"

scenarios. Target outcomes are defined in terms of the difference between "No" and "Yes", however, "correct" values will also be obtained (fortuitously) by using the difference between "Now and Yes", thus creating the illusion of a time-based effect. For example, when developing a new business, the "Now" scenarios (without the new business) will, in many cases serve as a surrogate for the "No" scenario.

2. **Now ≡ Yes**. Here, the funder seeks to *maintain* the "Now" scenario by undertaking a project. Without such intervention "Now" will, however, evolve into a different (undesirable) "No" scenario. In this situation, a time-based comparison gives rise to the illusion of performance levels being maintained, rather than changed. Consider a car maker who currently enjoys tariff protection in the domestic market, but who is faced with the removal of trade barriers as part of a free trade agreement. The firm's existing market share of 50% ("Now" scenario) is expected to fall to 45% as a result ("No" scenario). A marketing initiative is undertaken with the objective of retaining market share ("Yes" is also 50%). Here, the "Now" scenario can be used as a surrogate for "Yes". Because the definition of target outcomes is based on a comparison of "No" and "Yes", the target outcome of the venture would nevertheless, be correctly expressed as "*increased* market share by 5% (from 45 to 50%)".
3. **No ≡ Yes**. If the 2NY Map shows that the values in the "Yes" scenario are identical to (or worse than) those in the "No" scenario, it is clear that the project should not be funded.

A critical measurement issue now emerges, whereby, if a project is undertaken, the "Yes" scenario will be revealed and the "No" scenario will remain unknown. If the project is not undertaken, the reverse is true. How then are outcomes to be targeted? Setting values for target outcomes has to be done before a project investment decision is made and so the situation appears even more tenuous because, at that point, neither of the eventual "No" and "Yes" scenarios has been revealed. Anticipated outcomes must be set by *predicting* values for the variables that characterise the "No" and "Yes" scenarios. To deal with this issue, it is necessary to use some form of projection in a business case so that the values of the variables used to define target outcomes can be predicted for both the "No" and "Yes" scenarios. In most cases, the techniques required will be very obvious (such as simple "What-if" analysis), but in others, they may involve very sophisticated tools such as simulation analysis and statistical modelling.

It is clear from the 2NY map that the variables related to target outcomes must be measured *twice*: once when assessing the "Now" scenario and again when the project has ended. Measurement of the "Now" scenario, called "baselining", must be carried out before a decision is taken on funding a project so that reliable predictions can be made about the values of those same variables under the other two scenarios.

Measurements of the "Yes" scenario must then be taken when the project is over so that the success of the venture can be judged.

2.4.5 Baselining

The "Now" scenario in the 2NY map provides the *current* values for the selected variables-of-interest. *Future* values of those same performance variables are attached to both the "No" and "Yes" scenarios. It would normally be useful to know the "Now" values for each variable-of-interest before "Yes" and "No" values could be set, estimated or predicated. *Baselining* is the process of establishing the "Now" value of each variable-of-interest.

Baselining involves some sort of investigatory work which, as a process itself, will demand resources. The amount of work that baselining process requires will depend on the nature of the target outcome variables. At one extreme this may involve nothing more than having someone spend a little time checking on available sources of data, while at the other, a major study may be required, consuming significant labour and money. For various reasons, baselining is often not done, including two that bear some comment:

- Project participants are unaware of (or do not accept) the importance of measuring project performance.
- The cost or duration of baselining are deemed to be unacceptable for a particular project.

Except in rare instances, without "Now" values for the project's variables-of-interest, it is very difficult to determine effective "Yes" and "No" values for them. This in turn implies that target outcomes cannot be set. Because the achievement of target outcomes generates a project's benefits, this further implies that the benefits of the project remain unknown. If baselining is simply ignored, (and this could well be the situation for the majority of real-life projects) a funder is forced to assign notional values to all variables-of-interest under all three scenarios and make a qualitative judgement about the size of the "horizontal" difference in Fig. 2.4. If this *judged* value exceeds the "*true*" (but unknown) value, then the project could well be a worth while investment—but this can never be tested empirically. If we assume that, in such cases, funders tend to judge the value of such variables correctly, then most funded projects represent worthwhile investments—but this an empirical issue. A reasonable conjecture, however, is that decisions informed by baselining would tend to be more reliable than decisions without baselining.

This discussion begs the question "If it is to be done at all, when should baselining be undertaken?". Because the investment decision is taken in light of a business case, baselining should be done during initiation (see Chap. 9).

2.4.6 Naming Target Outcomes

The 2NY Map indicates that target outcomes are gauged by comparing the "No" and "Yes" scenarios and so valid names include "Increased/Decreased <name of measure>"—such as "Decreased waiting time for elective surgery". This word structure gives rise to "relative naming"—where the "Yes" value of the outcome variable is gauged relative to the "No" value. The value of an outcome variable can also be expressed in absolute terms by simply citing its value under the "Yes'" scenario (without reference to the "No" scenario).

While the choice of relative or absolute naming is arbitrary, there will, nevertheless, be cases where a relative name for a target outcome is less appropriate than the corresponding absolute name. To select the most appropriate of the two forms of name for a target outcome, the following word structures are suggested:

1. If, under the "No" scenario the variable for a target outcome does not exist at all (that is, it does not even have a zero value), then use the absolute word structure when naming a target outcome which takes the form of "New <outcome variable>"—such as "New vehicle sales".
2. Otherwise, use the relative form—"Decreased/Increased <outcome variable>"—such as "Increased vehicle sales".

Consider a project to establish a brand-new online retail service by a family who wants to increase their household disposable income. Assume also that they have no existing businesses of any form. Amongst the (many) alternative candidates for the target outcome being sought by the family is one with the relative form: "Increased business income" (of $55,000/year). Because there is no existing business income, this name, while completely correct, looks contrived. In such cases, an absolute form of name might be preferred—"New business income (of $55,000/year)".

Because target outcomes are associated with desirable end effects, it is tempting to express them in relative terms as an "improvement" in some variable. For example, in the case of the cross-city tunnel, a target outcome that captures the desirable end effect of decreased congestion might be described as "improved traffic flow". "Improved", however, should not be used in the title of a target outcome because it is unnecessarily vague. If the measurement of a variable indicates that it has improved, then the value of that variable has changed. If it has changed, it has either increased or decreased. We can conclude therefore that "improved" can always be replaced (more meaningfully) with "increased" or "decreased".

2.5 The Input-Transform-Outcome (ITO) Model

A fundamental question in the discipline of project management is "Why do we invest in projects?" If the answer is restricted to an IPO view of the world, it appears to be "To deliver outputs", but that simply begs the question "Yes, but why do we want those outputs?" The "real" answer clearly has to do with the end effects that those outputs can bring about.

We invest in projects to generate target outcomes because they will, in turn, generate a "flow of value" to certain stakeholders. Beneficiaries are defined as those who are better off from these flows of value as a consequence of a project achieving its target outcomes. At the same time, it is clear that outputs are a precursor to target outcomes and so the question arises: "How are outputs and target outcomes related?"

A number of conceptual devices that seek to explain this relationship have been proposed, including the Logic Model (Kellogg Foundation 2004) and the Outcome Profile(TM) (Walker and Nogeste 2008). While these models accept that inputs, processes, outputs and outcomes appear in that order, they offer no mechanism to explain what "causes" outcomes to appear. Similarly, PRINCE2 (OGC 2009), and Managing Successful Programmes (OGC 2011), popular proprietary project management methodologies, also highlight the importance of outcomes, but again, appears to ignore the transformation of outputs to outcomes. Because it does not recognise target outcomes, the IPO model is also clearly an inadequate representation of a project and so an alternative (based on an extension to the IPO view) is offered. Here we present the ITO model, which serves as our foundation theory.

2.5.1 *The Anatomy of the ITO Model*

A serious shortcoming in the IPO model as a representation of a project is that it makes no mention of target outcomes. To address this issue, we modify it by adding two elements (utilisation and target outcomes) to the three that are already included (inputs, process and outputs). The resulting five-element structure represents the ITO model of a project, as shown in Fig. 2.5. The model is so-named because it seeks to explain how Inputs on the left are Transformed into Outcomes on the right (Smyrk 1995).

The left-hand half of the ITO model is simply the IPO model, to which has been appended a utilisation mechanism and a flow of target outcomes. The original

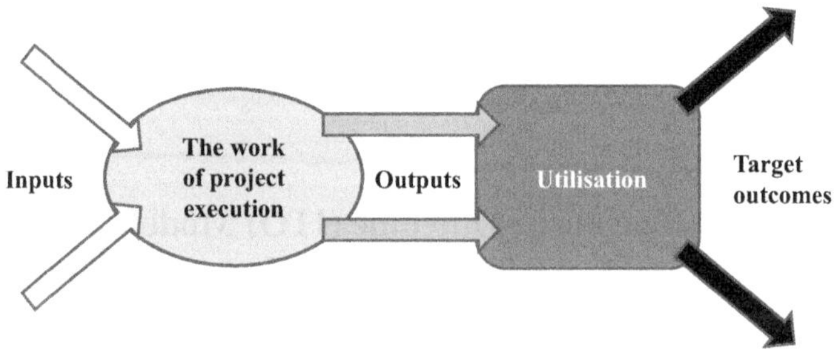

Fig. 2.5 The ITO model of a project

"left-to-right" chronology can now be extended. The project's outputs are eventually delivered to someone who then utilises them in a way that subsequently contributes to the generation of target outcomes. The entities who utilise a project's outputs in this fashion are called the project's customers. Project customers are not to be confused with: the organisation's customers, the project's beneficiaries or the project's funder (although, in certain cases, they can all be the same entity). A discussion on the distinction among these terms is presented in Box 2.6.

Practicalities

Box 2.6: An Important Distinction: Customers, Beneficiaries and the Client

We have taken a number of steps in our attempt to introduce rigour into the framework introduced here, including:

- Adoption of the ITO model as an organising theory. Much of our discussion emerges as extensions to and deductions from this theory.
- Establishment of a number of principles (which we use to fill a role akin to axioms in other disciplines).
- Assembly of an integrated glossary (see Appendix A).

Some elaborating comment is required about the last of these. Selection and adoption of appropriate terms for the elements of our framework involve trade-offs between:

- The need to reflect subtle nuances in the concepts we use.
- Creating new words unnecessarily.
- Acknowledgement of established usage of existing terms.

A particular illustration of the challenges faced here concerns our use of "customer". We distinguish between "a project's customer" and "an organisation's customer" in the following way:

- **A project's customer** utilises a project output in such a way that target outcomes are generated. If one were to display project customers in an ITO diagram, they would, of course, appear inside the utilisation rectangle.
- **An organisation's customer** is the entity to whom *operational* (as distinct from project) outputs are delivered. Commonly, when operational outputs are related to professional services, the term "client" is used instead of "customer". The delivery of an operational output frequently involves a commercial transaction of some kind. This use of the word "customer" accords with its commonly accepted meaning.

Because the same entity *frequently* fills both roles, some of the existing literature (mistakenly) assumes that they are *always* the same. To distinguish these roles we will, where necessary, qualify "customer" with the words "project" or "organisation".

Other distinctions need to be highlighted as well. Two are particularly important because they are surrounded by considerable confusion in the literature.

- **A project's client**. A project manager is commissioned as supplier (of project outputs) by the funder (or the project owner on behalf of the funder). Whoever commissions the project manager can, in a sense, be regarded as his/her "client".
- **A project's beneficiary**. An entity who enjoys a "flow of value" arising from achievement of target outcomes from the project. Beneficiaries can be project customers, but in general, this is not the case. We do not refer to beneficiaries as "customers".

In the left-hand (IPO) part of the ITO model, there is *strong causality* between the work of the project and its outputs. This means not only that the outputs can exist only if the process is executed, but also that, if the process is executed, the outputs will be created. Consider, for example, a blow-moulding process to produce a plastic pallet. Each time the process is executed (in accordance with its script) we get a plastic pallet. Furthermore, if we see one of these pallets, we know that a blow-moulding process has been executed.

By way of contrast, in the right-hand part of Fig. 2.5, the link between outputs and target outcomes is an example of *weak causality* whereby the delivery of outputs does not guarantee the generation of target outcomes.

An illustration of the ITO model for a road project is provided in Box 2.7.

Illustrative example

Box 2.7: An ITO Representation of a Road Improvement Project

The following project provides illustrations of the ITO terminology. A project is funded by the state roads authority to reduce the interruptions to traffic flow at a rail level crossing. A valid ITO model for this project includes:

1. **Inputs**—funds (for raw materials and machinery) and labour (measured as person-hours).
2. **Process**—build an overpass.
3. **Outputs**—overpass, removed level crossing and signage.
4. **Utilisation**—commuters driving across the overpass.
5. **Target outcomes**—decreased travel time (between two reference points—one on each side of the rail line) and increased commuter satisfaction.

The ITO model for this project is shown in Fig. 2.6.

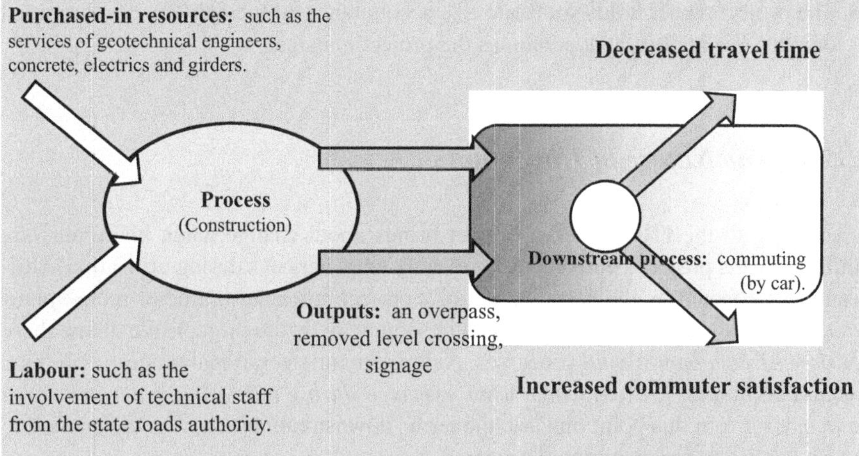

Fig. 2.6 An ITO model of a road improvement project

Because of strong causality, we treat processes in general and project execution in particular as controllable. This means that given enough time and resources, we can produce any technically feasible output and are, therefore, willing to guarantee (eventual) production of that output. Consider a project to replace an ageing narrow steel truss bridge with a wider new concrete girder design. The bridge runs across a river separating a heavy industrial estate from a harbour that handles both general cargo and containers. A contractor for such an exercise would normally be quite comfortable guaranteeing the replacement because the work involved can be controlled to a high degree. By way of contrast, because of weak causality, in general, we cannot guarantee the eventual generation of target outcomes. In the case of the bridge replacement, assume that the target outcome is reduced travel time, with a target threshold of 25%. Many factors could prevent this result from being achieved (such as an unexpected increase in the number of wide-load trucks that were previously banned from using the steel bridge).

2.5.2 Accountability in the ITO Model

Foreshadowing the coverage of Chap. 4, the ITO model has implications for project governance. It would appear appropriate to make someone accountable for each of the two types of *result* from a project (outputs and target outcomes). Here, in addition to the funder, we introduce another two of a project's key players: the project manager and the project owner who hold the following accountabilities:

- The funder holds someone accountable for the realisation of target outcomes. We identify that person as the project owner.
- The project owner holds someone else accountable for the delivery of committed outputs. We identify that person as the project manager.

2.5.3 The Nature of Utilisation

According to the ITO model, a project brings about change when its outputs are utilised by the project's customers. So exactly what happens during utilisation? Utilisation is effected by the participation of a project customer in one or more *operational processes* that have been altered in some way by the project. We define these as *downstream operational* processes. A downstream operational process is defined as *an operational process which when executed during utilisation generates target outcomes*. From this point onward, the term "downstream process" is understood to mean "downstream operational process".

We propose a *principle of duality* whereby the utilisation of a project's outputs by its customers *is equivalent to* the execution of downstream processes by project customers. According to this principle, *the target outcomes from a project can always be expressed in terms of the changes in particular variables observed during the execution of specific downstream operational processes*. In the language of the process engineering discipline, these are called process metrics—used to gauge the performance of a process. Process metrics are of two kinds: those that can be measured for each execution of the operational process and those that can be derived only from a sequence of executions of the operational process. Process execution time for a procurement process is an example of the first of these, while accident rates for a random breath testing initiative is an example of the second. In a typical project to improve the performance of a procurement process (see Sect. 2.4.3) one such metric might be "Procurement cost"—obtained as the value of all the resources consumed whenever the process (procurement) is executed.

Figure 2.6 shows how the execution of a downstream process relates to the utilisation mechanism. Note that a process qualifies as a project output—as explained in Sect. 2.4.5.

In Fig. 2.7, the outputs from the project are delivered into some form of operational environment. A project's key players and its customers see utilisation from very different standpoints. As far as key players in the project are concerned, following project execution, the project's customers are simply utilising the outputs that have been delivered, and so, any change in the measured values of target outcomes is a consequence of that utilisation. For example, the opening of a wider vehicular bridge has commuters utilising the bridge to reduce travel time. However, as far as the project's customers (road users) are concerned, they are merely executing the regular (operational) process of commuting between home and the central business district. Here, drivers are primarily interested in a particular metric associated with the commuting process–namely, trip-time. The availability of the bridge now

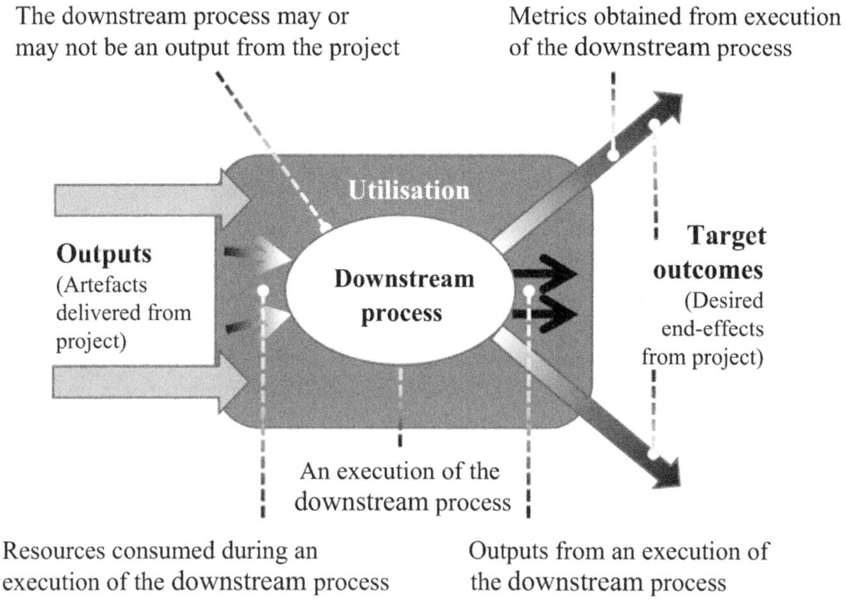

Fig. 2.7 The utilisation mechanism for a single execution of a downstream operational process

allows them to undertake the downstream process (commuting) differently and so the trip-time metric changes. From the perspective of a project customer, new outputs now enable an operational process to be executed differently and, as a result, certain metrics associated with that operational process change. In the terminology of the process engineering discipline, a project customer can be viewed as a "process agent"—someone who participates in (or facilitates) the execution of an operational process—a view that aligns with the term "user" in common parlance.

Figure 2.8 shows the most general situation, whereby utilisation involves an indeterminate count of executions of the downstream process into the distant future. In the example of the Cross-City Tunnel (Sect. 2.4.4), the outcome "Reduced noise levels" may be generated every day into the future for as long as the tunnel remains in operation. This would be true for many projects, but in some instances, utilisation involves a limited count of executions of the downstream process. Take the case of a 5-day exhibition of rare paintings in a provincial city for which the target outcome is "increased awareness of the services provided by the national museum". Here project execution involves all the work undertaken up until the official opening. Utilisation then takes place during the exhibition itself (where visitors look at the items on display)—with each visit representing an execution of the "viewing" downstream process. Because the exhibition is not permanent, utilisation is time-limited—as will be the flow of target outcomes. In certain projects, utilisation may allow only *one* execution of the associated downstream process—as in the case of a rock concert–where project execution involves all the work undertaken up until the

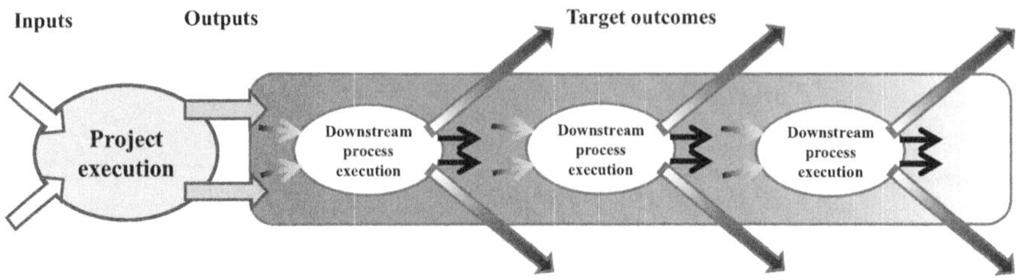

Fig. 2.8 The utilisation mechanism for multiple executions of the downstream process

gates open, while utilisation is the conduct of the concert itself. In this case, there is only one execution of a single (albeit substantial) downstream process involving large numbers of concert-goers as process agents.

2.5.4 The Impact of Projects on Operational Processes

Operational processes can be impacted by projects in any of three ways:

1. **The creation of a new downstream process** (where none existed before). Consider a restaurant that decides to establish a new home delivery service. In this case, the new business would involve creation of multiple new downstream processes (such as *accept orders by phone*, *package food for delivery* and *despatch courier with packaged meal*).
2. **The moderation of an existing operational process**. Here there is already an operational process which the project impacts either directly or indirectly. Moderation can take either of two forms: *modification* or *influence*. A project impacts an operational process *directly* when the project team *modifies* an existing operational process and that modified process emerges as an output from the project. The project to re-engineer a company's procurement processes (see Sect. 2.4.3), provides an example of process modification. By way of contrast, the construction of roundabouts, traffic islands and speed humps to reduce speed-related accidents on a suburban street merely influences the existing operational (driving) process so that it takes longer to execute.
3. **An operational process is terminated**. BHP is one of the world's largest mining companies. Throughout the twentieth century, it was also a steel producer. Because international competition eventually made this business uneconomic, various steps were taken to stem the losses from steel-making operations—including the closure of an integrated iron and steel plant in Newcastle (an Australian regional city). The project to close the plant (and later demolish it) resulted in the termination of a large number of operational processes—with a corresponding reduction in operating costs.

When an operational process is terminated, its metrics necessarily fall to zero—as was the case for the costs associated with the operation of the (now demolished) BHP steel plant. The achievement of a zero target for a downstream process metric (or, looked at another way, the achievement of a zero target for an outcome) happens automatically. In all other cases, the generation of target outcomes requires that one or more downstream processes be executed by project customers. Those outcomes for which targets are achieved automatically (without utilisation of any project outputs) are said to be "natural" outcomes.

A newly created or modified downstream process is called *endogenous* (because it emerges from within the project). An existing operational process which is merely influenced by the project is called *exogenous*. In effect, an endogenous downstream process is delivered into the operational environment as an output from within the project, while an exogenous downstream process is delivered into the operational environment from outside the project.

It should be noted, that the destruction of an artefact or natural object is not necessarily associated with the termination of an operational process—as is the case with Ripple Rock (Box 2.8), where a dangerous reef was destroyed so that ongoing navigation (an exogenous process) was made safer.

Illustrative example

Box 2.8: Ripple Rock: the moderation of an operational process through the elimination of a natural object

Ripple Rock was a dangerous undersea pinnacle (rising to within 3 m of the water's surface in the Seymour Narrows near Campbell River, British Columbia in Canada) (Wright et al. 1958). Fast currents sweeping over the reef created extraordinarily chaotic and dangerous sailing conditions, which, over the years, are reported to have sunk more than 120 vessels with the loss of 114 lives. After a number of unsuccessful attempts to deal with the problem, in 1958 a project was undertaken to remove the reef. This was achieved by drilling down through nearby Maude Island, horizontally under the bay and then vertically up into Ripple Rock itself. There, a network of "coyote" tunnels was filled with over 1,270 tonnes of high explosive. Ripple Rock was then destroyed in, what was at the time, the largest ever non-nuclear peacetime blast.

An example of the valid outcomes and outputs that could be used to define this project are:

- Target outcomes: reduced loss of vessels and lives in the Seymour Narrows.
- Output: an eliminated reef (in the terminology introduced earlier, a "null").

In this case, the downstream process takes the form of navigating the nearby shipping channel. The loss of lives and ships is an example of a process metric, which, in this case, would be obtained by taking a sample of passages and measuring such variables as: count of deaths and value of ships damaged and lost. (While a single observation of such metrics from a single passage would be meaningless, observations over an extended period of time would make sense.)

After removal of Ripple Rock, the process of navigating the channel was executed in a different environment and so those same metrics revealed significant reductions in deaths, ships foundering and damage.

According to the categories of operational process introduced above, navigating the channel is *exogenous* because it has been merely influenced (indirectly moderated) by the project.

It should be noted that an operational *process* qualifies as a project *output* because initiatives directed at the creation or re-engineering of a downstream process do so through the design, development and delivery of a model of the process. Such a model always takes the form of an artefact—for example, as a Business Process Modelling Notation (BPMN) diagram. Furthermore, because it can be represented by such an artefact, an existing process can serve as the operand for a project—thus falling under the "modify" category in Table 2.1.

The project to re-engineer a company's procurement processes (see Sect. 2.4.3) provides an example of an endogenous downstream process—where a re-engineered version of an existing procurement process is delivered as a project output.

As a further example of an exogenous downstream process, consider a manufacturer of pharmaceutical tableting machines (Tabco Inc.) undertaking a project to design a new product and build a production line for its assembly. Assume that Tabco's target outcome is "increased sales revenue", and that the project's outputs include: a design and prototype of the new machine, a new manufacturing process and a new factory but which exclude the (existing) sales process—the script of which remains completely untouched. In what sense then might the sales process be influenced by Tabco's new product project if the sales process script is not changed? In this case, "influence" would relate to such process metrics as: the average value of a sale, the frequency of sales and the gross profit on a sale—none of which require a change in the script for the sales process. This implies that the downstream (sales) process is exogenous.

Summary

In this chapter, we have examined the execution of a project as a special case of a process—tied to an investment decision. A project is funded with the specific intention of generating clearly articulated target outcomes that will lead to a flow of benefits to particular stakeholders. All of the established descriptive and analytical devices that relate to processes—such as the IPO model—apply equally to projects, but, because conventional approaches do not recognise target outcomes, we have proposed a new tool—the ITO model—to address this shortcoming. Amongst other things, the ITO model exposes the sharp difference between outputs and outcomes—as well as offering an explanation of how outcomes are generated by a project. A key element of the ITO model is the concept of utilisation where project customers generate a project's target outcomes by utilising its outputs. A principle of duality emerges from this discussion. The target outcomes from a project are identical to certain process metrics obtained from an operational process that has been created or moderated by the project. When viewed from the perspective of key players in a project, target outcomes are generated through the utilisation of outputs by project customers. From their perspective however, project customers are engaged in the execution of a downstream operational process which results in observable changes in the values of various process metrics. *A fundamental implication of the principle of duality is that all projects are, in effect, initiatives to enhance operational performance through process engineering of one kind or another*. In the next chapter, we discuss how the ITO model impacts the structure of a project and its life cycle.

Chapter 3
The Structure of a Project

Abstract The *project* environment is usefully distinguished from the *operational* environment (where routine operational processes are executed). The project environment is characterised by three structures imposed on this work so that it can be managed systematically. In the first of these, the life of a project is broken into four *global phases*. The second offers an anatomy for a project in which all the core information about a project is classified into seven *elements*. The third separates the work on a project into two *layers*: one related to the production of the project's outputs and the other involving the management of that activity.

3.1 Project Global Phases

Undertaking a project involves four *strictly sequential* major "blocks of work". Each of these *global phases* consume resources and result in an output:

1. **Initiation**: which seeks to answer the question: "Is this project a worthwhile investment?". The answer leads to in principle acceptance (or rejection) of a business case, which contains all the information required for a reliable funding decision. Initiation is covered in Chap. 9.
2. **Planning**: in which a script for the work to produce the project's deliverables is mapped out in detail. The primary output from this global phase is a *project plan* on which approval is given to begin substantive work. Although the project owner is accountable to the funder for the entire project, the actual planning is delegated to, and carried out by, the project manager. Planning is covered in Chap. 10.
3. **Execution**: when the project's outputs are produced, delivered and implemented. All projects are completely dominated by execution because it takes most of the overall elapsed time and consumes most of the available resources. During execution, the project manager is responsible for the conduct of all the work involved—as detailed in the project plan. Execution is covered in Chap. 11.
4. **Outcomes realisation**: when, (through a programme of appropriate intervention), attempts are made to secure the flow of target outcomes. During this global phase, the project owner is responsible for ensuring project outputs are utilised to

O. Zwikael and J. R. Smyrk, *Project Management*,
https://doi.org/10.1007/978-3-030-03174-9_3

generate the target outcomes. The skills required to guide and facilitate outcomes realisation are similar to those that are necessary for the effective management of execution. In most cases, the person who filled the role of project manager will be retained to administer this global phase on behalf of the project owner. (It should be noted that the *role* continues, even though accountabilities come to an end when output delivery is complete). Outcomes realisation is covered in Chap. 12.

Because downstream operational processes are executed during this global phase, it actually forms part of the operational environment, but with a critical difference. Outcomes realisation has a fixed duration, during which accountability for the realisation of target outcomes lies with the project owner (Zwikael and Smyrk 2015). If downstream processes continue to be executed beyond that date, accountability for target outcomes transfers to a functional or operational executive.

In contrast to the conventional view (e.g. PMI 2017; Kerzner 2013), we propose that the span of a project's life does not end with delivery of outputs (at the end of execution), but that it continues until target outcomes are secured (Zwikael and Meredith 2018). Given the importance of outcomes in projects, outcomes realisation has been introduced as an additional global phase. Another difference with existing models is our treatment of closeout: the ITO model implies that closeout is required after both the delivery of outputs and securing of outcomes. As a result, closeout processes are included in both of the last two global phases rather than being treated as stand-alone activities.

Table 3.1 shows a proposed generic structure of global phases for a project. This represents a high-level description of the work that is common to all projects.

As the hierarchy in Table 3.1 suggests, these global phases are broken out in turn into a number of processes that are also common to all projects. (Note that, some of the processes within a global phase may be executed in parallel).

The process ellipse in the ITO model is concerned only with the work of producing a project's committed outputs. Because the work undertaken during initiation and planning is concerned only with preparing for the production of those outputs (rather than their actual production) they do not correspond with elements of the ITO model. Execution and outcomes realisation, however, are represented explicitly in the ITO model—as highlighted in Fig. 3.1. Further, each of the four global phases is described in the following discussion.

3.1.1 Project Initiation

Initiation has four distinct processes: identification (where the foundations of the project are established), definition (where values are set for certain critical parameters, analysis (where estimates are made of various derived parameters) and packaging (where the business case is assembled).

Table 3.1 A generic structure for a project's global phases and related processes

The global phase structure of a project
1. Initiation
• Identification • Definition • Analysis • Packaging (of the business case)
2. Planning
• Project plan preparation • Packaging (of the project plan)
3. Execution
• Project environment set-up • Output production • Outputs closeout • Execution wind up
4. Outcomes realisation
• Facilitation of output utilisation • Outcomes closeout • Handover to operations

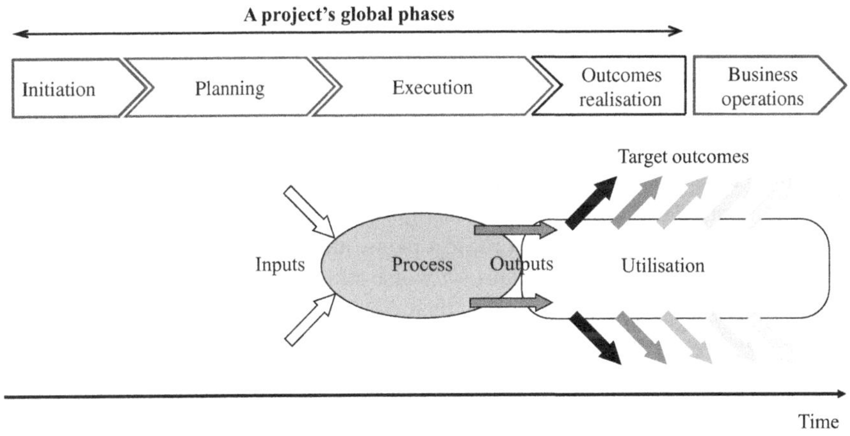

Fig. 3.1 A project's four global phases mapped onto the ITO model

It is important to note that the purpose of a business case is to support a reliable funding decision rather than to sell the project or justify it. A confident decision not to fund would suggest that the business case had served its purpose successfully.

At its inception, each project has a "champion"—someone who is charged by the funder with responsibility for assembling the business case. While the champion drives initiation, it is the funder who will make the final decision to accept or reject the business case. As will often be the case, when the champion does not have the capacity or the capability to assemble the business case him/herself, the work is assigned to someone who is appropriately skilled. That person is clearly a candidate to fill the role of project manager and might well be described as the "project manager designate". In that event, it should be noted that the business case is still owned by the champion, who will eventually present it to the funder. In assembling the document, the project manager designate is simply supporting the champion and at no time becomes accountable for its tabling or acceptance.

When the funder accepts a business case, he/she will normally commission someone as his/her agent to assume responsibility for the project overall. That person becomes the project's owner, accountable to the funder for realising the business case and hence eventually securing the flow of target outcomes. There is nothing to prevent the funder from acting as owner. In most situations, however, funders are responsible for such a large portfolio of projects that it is not feasible for them to give each one the close attention it demands, and so it becomes necessary to appoint someone else as project owner. The champion is clearly a candidate to fill that role. At this point, the project owner will usually engage a project manager. See Chap. 9 for a detailed discussion on initiation.

3.1.2 Project Planning

During this global phase, the project manager develops a detailed project plan. A project plan is essentially a script—describing *how* a project's outputs are to be delivered and implemented—with considerable supporting detail about how the project is to be approached. The preparation of this document usually takes a significant amount of time to complete and demands the allocation of appropriate resources. It is worth noting that while the construction sector tends to recognise the need for high quality, complete and robust project plans, many business projects are started with unreliable and incomplete plans (or worse still, with no plans at all!). Once the project plan has been developed, it is then considered for approval by the project owner (if necessary, in consultation with the funder). See Chap. 10 for a detailed discussion on planning.

Relative to initiation, planning (intentionally) produces a large amount of additional information about the project. This additional information may result in changes to some parameters appearing in the business case.

3.1.3 Project Execution

During this global phase, the project's outputs are created by undertaking the work described during planning. Before a start can be made on producing the project's outputs, some preparation is required, in the form of a set-up activity. Here, the project environment is established by marshalling the team, implementing the governance model and in some cases, by assembling the project's own infrastructure (such as offices, vehicles and systems). Once this has been done, then work on the actual production of outputs can begin. Because project plans are usually assembled with information of widely varying completeness and reliability, they require continuous ad hoc modification and refinement as the work unfolds. Control and management of that type of change is based on a separate (but closely connected) stream of administrative activity, mainly concerned with monitoring, tracking, reporting and guiding.

Following formal outputs closeout (see Chap. 11), execution concludes when various elements of the project's governance model are stood down.

3.1.4 Outcomes Realisation

This global phase represents a significant extension to the work that has traditionally defined "the project". In terms of the ITO model, target outcomes are generated by utilisation of outputs. Outcomes realisation (for which the project owner is accountable) is that portion of output utilisation that occurs up until the flow of target outcomes has been secured. Beyond that point, the generation of target outcomes takes place as part of regular business operations (for which a line manager is typically accountable). Outcomes realisation ends (and the project concludes) when, within an agreed timeframe, flows of desirable outcomes stabilise below, at or above the target thresholds that have been set. Outcomes realisation involves three major processes: facilitation, outcomes closeout and handover. During facilitation, an attempt will be made to influence the generation of target outcomes with an appropriate programme of intervention to enhance output utilisation.

Outcomes closeout affords key stakeholders the opportunity to review outcomes achievement and make judgements about the success of the project. Output from this process is an outcome closure report, which analyses the extent to which the project's objectives have been achieved.

Outcomes realisation ends with release of some form of "end-of-project declaration" confirming not only that target outcomes are secured, but also that they are likely to continue flowing into the future. It is important to note that the project manager and team discharge their responsibilities when outputs are implemented. Some members of the team (and even the person who acted as project manager) may remain during the outcomes realisation phase, but their role is now quite different, simply supporting the project owner and customers. See Chap. 12 for a detailed discussion of the outcomes realisation global phase.

3.1.5 Global Phases and Accountabilities

The foregoing discussion suggests that primary project accountabilities shift throughout the life of the project, as is described in Fig. 3.2.

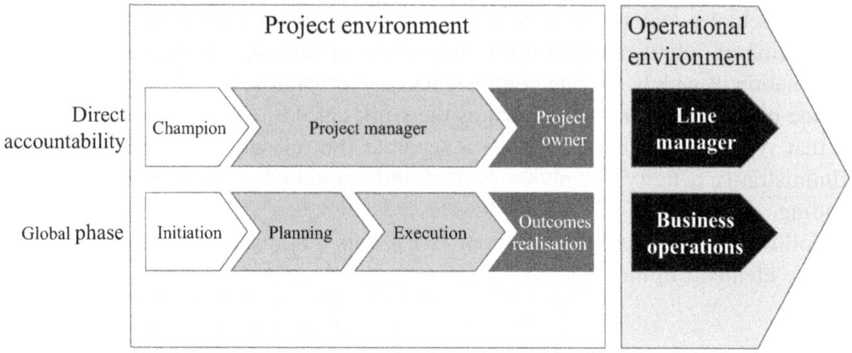

Fig. 3.2 Direct accountabilities during a project's life

Figure 3.2 shows that the project manager is only one of the several stakeholders who hold accountabilities in different global phases. This makes it absolutely clear that a project is not a "one-person show" featuring the project manager; instead, it is a collaborative effort involving many key players with various related accountabilities. For example the project champion (holds an accountability for project initiation and the development of the business case), project manager (for planning, execution and output delivery), project owner (for outcomes realisation management and securing a future flow of target outcomes) and line (or functional) manager (for any ongoing generation of outcomes).

3.1.6 Staged Projects

The special case of staged projects is worthy of discussion. Staging arises when a series of go/no go decisions must be made as the venture unfolds. A sequence of such decision effectively breaks the project up into serially dependent stages where the start of work on one stage is dependent on the completion of another. R&D projects are typically of this kind—although sequential dependency can arise in other situations as well. The progress of a new drug from research lab to marketplace involves strictly serial stages—based on a protocol established by the U.S. Food and Drug Administration (FDA).

An example reveals why scoping such a project must be approached iteratively. An independent research lab ("IRL") has developed a new drug for asthma. It approaches a large pharmaceutical company ("LPC") with a proposal to manufacture, market and distribute the product under licence. A naive response to this approach by LPC would see a project with a target outcome of increased revenue and outputs including a manufacturing plant, marketing campaign and distribution network. Clearly it would make no sense for the LPC Board to fund such a project because of the numerous unknowns: Can we get FDA approval? Is IRL's production pilot scalable? Does the drug work? Might it cannibalise an existing product? Is it consistent with long term business strategy? It would appear reasonable, therefore, to break the exercise into well-defined sequential chunks—with an internal review of IRLs research results as a first step.

If we now apply the process trichotomy introduced in Chap. 2, the review of IRLs research must be itself undertaken as a (preliminary) *project*. In ITO terms, the dominant output from such a project is a report of some kind. An appropriate target outcome would be labelled something like: "increased confidence in the decision whether to proceed with stage #2" (whatever form that was to take). If, after the tabling of the stage #1 report, the board was able to make a confident decision about progress to the next stage, then this outcome has been achieved. (Note that a confident decision to abandon the initiative is just as indicative of success as a confident decision to proceed).

We now have framework for a staged project in which each stage itself is also conducted as a project. Except for the last one, each stage has a single generic target outcome along the lines of "Increased reliability of a decision about funding the next stage". The output from these intermediate stages would take the form of documents, certificates and reports, prototypes and pilots. In addition, each stage would also assemble the business case for the next stage. Clearly, the overall staged project is driven by the target outcomes attached to the last stage. If, for example, FDA approval is not obtained, then the Company has certainly achieved the target outcome set for this stage (of making a reliable decision to abandon the exercise), despite the fact that it is denied achievement of any objectives it might have set for the final stage. Figure 3.3 illustrates the structure of a staged project.

Sequential dependency does not necessarily result in a staged project. Take the case of two projects A and B where: A (undertaken by a port authority) seeks to increase the throughput of a port by constructing additional shipping berths and B (undertaken by a local alumina refinery) seeks to reduce operating costs by building a high-capacity ship-unloader for the bauxite produced at its nearby mine. The ship-unloader requires new berths which the refinery must either build itself or use those being proposed by the port authority. Clearly, there is an incentive for the refinery to adopt the second strategy—in which case, although the start date for B is not related

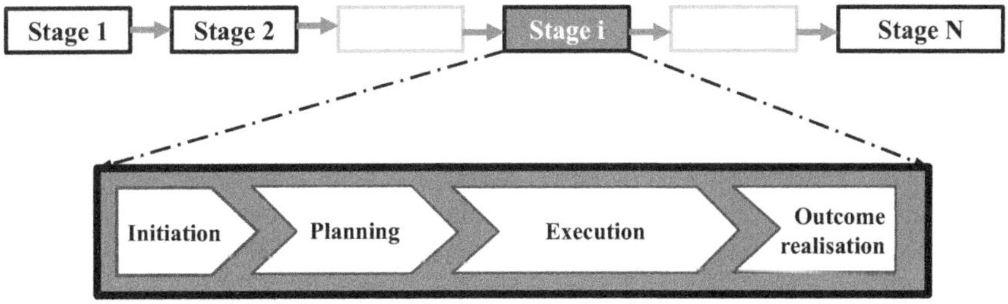

Fig. 3.3 The structure of a staged project

to A, it cannot be completed until A is finished. Although *sequentially dependent*, for two reasons, A and B do not constitute a *staged* project:

- A start to "B" is in no way dependent on achievement of the *outcome* of "A".
- Both "A" and "B" make sense as stand-alone investments.

While they are not *stages* of some larger overarching project, "A" and "B" do, *interact*. More is said about the management of interacting projects in Chap. 4.

3.2 The Elements of Project Management

In addition to progressing through global phases (a time-based perspective), project management can be shown as made up of elements representing a structural view.

All project methodologies employ a taxonomy of core elements which frame their management guidelines. For example, PRINCE2 (OGC 2009) recognises: organisation, plans, controls, business case, risk, quality, configuration management and change control. The PMBOK (PMI 2017) identifies ten knowledge areas: integration, scope, time, cost, quality, human resources, communications, risk, procurement and stakeholders. If it is to be effective, at its highest level, a taxonomy of elements should be exhaustive and exclusive while allowing for the addition of others that eventually provide a "slot" for every concept used in the associated discourse. Unlike other frameworks, where the classification of project elements appears to arise from a largely ad hoc grouping of "accepted practices" (a bottom-up approach), we suggest an alternative top-down approach using a seven-elements taxonomy of project concepts:

1. **Scope**. According to the ITO model, the generation of target outcomes requires that particular outputs be delivered with specific features, and then it is utilised by the project's customers. A project's outputs must therefore be identified, and then defined by a list of *fitness-for-purpose features* (which, in turn, make them "utilisable" by project customers). The quality of an output is reflected in its list of fitness-for-purpose features. The completeness of an output's definition determines its "fitness-for-purpose". As the "fitness-for-purpose" of an output (i.e. its quality) falls, so does the effectiveness of its utilisation by a project customer (and its eventual contribution to the generation of target outcomes). We are thus led to an important principle of project management: a project is scoped if and only if its outputs are set. A project's outputs are set when two conditions are met: (a) They are all listed and (b) Each is defined by a list of critical fitness-for-purpose features. Scoping is central to project management because it rigorously establishes a reliable list of outputs, without which target outcomes cannot be generated. Once outputs are set, it then becomes possible not only to plan the project (and in particular to describe the work required for their production) but also to manage any changes in scope during project execution. Scope is covered in Sect. 5.2. Accordingly, scope is included as a structural element of our framework.

2. **Stakeholders**. Some entities can influence the way projects unfold, while others are impacted by the project. Because stakeholders, by definition, have an interest in a project, the nature of this interest should be analysed and managed, and so "stakeholders" are included as a structural element. Stakeholder management is covered in Chap. 5.
3. **Governance**. A project involves various key players, each of whom fills a specific role throughout the life of the project. Effective management of these key players requires an overarching governance model. The standing organisational arrangements in most organisations are based on functional roles and thus do not accommodate projects very well. It is necessary, therefore, to establish special organisational arrangements for projects, and so "governance" is concerned with the organisational aspects of projects and is, accordingly, included as a structural element. Because projects are not repeated, the organisational arrangements that surround them represent a temporary extension to the standing structure of the funding organisation. The project's organisational structure is dismantled when the project ends (on securing target outcomes). Governance is covered in Chap. 4.
4. **Workplan**. The work associated with a project has to be analysed so that reliable dates can be attached to critical points during execution, particularly those relating to delivery of outputs and realisation of outcomes. Project execution requires a "script" if it is to be executed successfully. An effective script for a project can take any of a number of forms, the most useful of which involve a Work Breakdown Structure (WBS) and Gantt chart (both introduced in Chap. 9), as well as a schedule of milestones. The need for an executable model of the work in a project is reflected in our recognition of schedules as a structural element. Scheduling is covered in Chap. 7.
5. **Resources**. Outputs require work and work demands resources. The purchase and deployment of resources generate project costs—which appear as one of the three variables in a project's "equation of worth". Management of resources—especially during execution—merits inclusion in the list of project elements. Resource planning is discussed in Chap. 10.
6. **Risks**. Projects are subject to all sorts of uncertainty which must be managed. Project uncertainty is formalised as risk. Understanding the level of risk to which a project is exposed is a precondition for accepting the business case. A project portfolio can be viewed as a list of investment opportunities, ranked according to their attractiveness. A project's attractiveness is determined by its worth and "riskiness"—as discussed in Chap. 7. Riskiness is a gauge of a project's exposure to falling short of its target worth. A framework for the management of risk is, therefore included as a structural element. Risk management is covered in Chap. 6
7. **Issues**. The project environment is also subject to other sorts of developments that, while not necessarily representing a threat to success, demand a response. We label these events as "issues". Projects face a continuous barrage of spontaneous events, most of which are low-scale in nature. In some cases, these events, and the reactions to them by key players, result in a progressive evolution of

the project environment. A framework for the management of issues therefore merits inclusion as a structural element. The issues are covered in Chap. 6.

3.3 The Layers of Work in a Project

3.3.1 Above- and Below-the-Line Work

Work undertaken in a project is of two distinct kinds. The first (and most obvious) is that required for the production, implementation and delivery of outputs. We call this *below-the-line* work. Below-the-line work is confined to the third global phase (execution). The project's deliverables may therefore be termed "below-the-line outputs".

The work required in the third global phase to develop the "below-the-line outputs" also requires ongoing management. We call this *above-the-line* work. Above-the-line work gives rise to above-the-line outputs, such as schedules of milestones, project progress reports and various registers. Above-the-line work is associated with maintenance of the seven structural elements discussed above.

From the literature

Box 3.1: Above- and Below-the-Line

The terms "above-the-line" and "below-the-line" are used (with unrelated meanings) in marketing, accounting and contract management.

In the marketing profession, the terms are used to classify sales promotion activity. Above-the-line refers typically to mass media advertising. Below-the-line activity is transaction-oriented (such as price discounting, coupons and point-of-sale promotion). The distinction is important to advertising agencies where below-the-line promotions (under normal conditions) earn no commission, while above-the-line activity does.

The accounting profession uses "above-the-line" to mean cost-of-good-sold (or equivalent), while "below-the-line" refers to operating expenses. The contract management profession uses the terms to classify the costs incurred in contracts. There, below-the-line refers to those factors that can impact the value of a contract after the basis of an estimate has been fixed (and agreed). Below-the-line items include those factors that lie beyond a contractor's control, such as escalation and exchange rate variation. By way of contrast, above-the-line refers to factors that, under normal conditions, have no implications for the value of the contract.

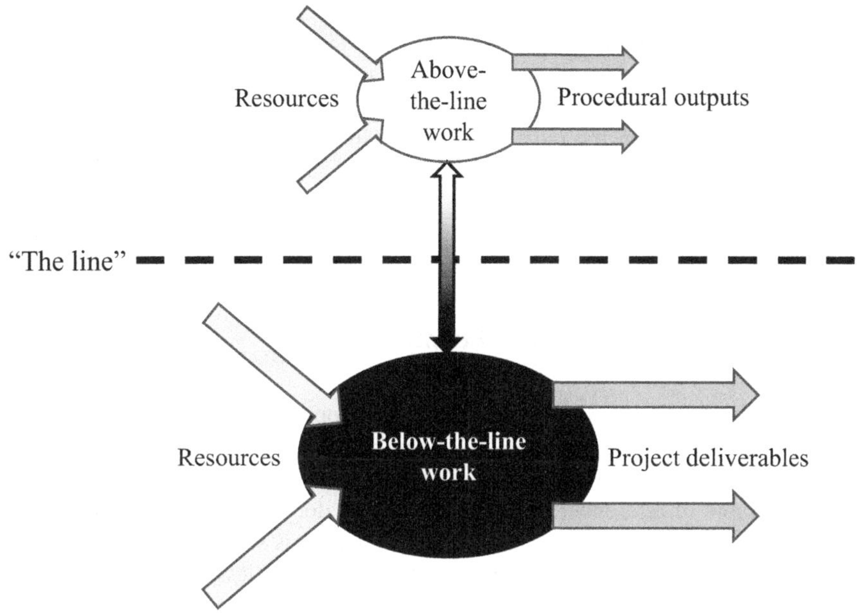

Fig. 3.4 The relationship between "above-the-line" and "below-the-line" work undertaken during project execution

Figure 3.4 shows how the two classes of work are related. Below-the-line work is concerned with the project's deliverables. Above-the-line work is concerned with controlling the project's below-the-line work. These two concepts are further discussed in Boxes 3.1 and 3.2. The term "procedural outputs" in the figure refers to items such as the business case, the project plan, and the various documents used to track the progress of the project.

Practicalities

Box 3.2: How Much Project Load Arises from Above-the-Line Work?
A common rule of thumb (for which there is some supporting anecdotal evidence) is that the ratio of internal labour consumed by above-the-line work to the total internal labour consumed by the whole project is about 15%. (Note that, this ratio involves only the labour deployed or hired under contract by the organisation undertaking the work, it excludes the cost of purchased products).

If, for example, a project is expected to consume about 15,000 person-hours of company staff and contractor time, then another 2,250 person-hours (or so) will be consumed in all above-the-line work across the life of the initiative.

This now raises a question "What happens to this ratio as the project size increases?". A priori, a plausible argument can be advanced suggesting that diseconomies of complexity grow faster than economies of scale. If that conjecture is right, then the ratio of above-the-line effort to total project effort would rise with increasing project size.

3.3.2 A Project's Baseline Documents

Above-the-line work is associated with the development of the many documents that are necessary for the management of a project. A project's *baseline* documents (the business case and project plan) are cases in point. Together, they establish the project's frame of reference. In addition, baseline documents guide all significant aspects of a project's processes and decisions, as well as setting out the roles of all key players. Nothing of significance should be effected unless, it is consistent with the project's baseline documents. The business case and project plan are discussed comprehensively in Chaps. 9 and 10 respectively.

Summary

By imposing a common structure of four global phases, seven elements and two layers on projects, we are now able to establish a universal frame of reference which can then be used to systematically describe, analyse and model them. Such a model can then be used as the foundation for management of all the work involved over the life of a project. The next chapter explores the governance of projects and programmes.

Chapter 4
Project and Programme Governance

Abstract To manage projects and programmes effectively, an *organisational model* needs to guide the working arrangements for all those who are involved. Project governance is concerned with the identification of the key players and descriptions of their roles in the project. Similarly, a governance structure is also required for programmes, so that related projects are effectively coordinated. This chapter first discusses project governance, followed by programme governance.

4.1 Project Governance

A project usually involves a significant amount of work undertaken by a large number of people, hence some sort of management and administrative framework is required to organise all of those who are involved. Project governance seeks to create a structure within which all those involved in the project can fulfil their roles effectively. Such a structure is called a project governance model. An organisation's standing management arrangements usually focus on *operational* processes, hence requiring accountabilities and reporting lines that are normally inappropriate for *project* governance.

4.1.1 Overview of Project Governance

One of the major determinants of project success is an effective project governance structure (Musawir et al. 2017). Given the temporary nature of projects, each one requires a unique governance structure which, while distinct from the relatively stable standing structures of the participating organisations, must nevertheless, coexist with them. The assignment of accountabilities to certain entities in the project governance model is important, because it helps bridge the gap between the expectations of a role and the way that role is filled—by attaching sanctions and rewards to levels of performance (Zwikael and Smyrk 2015).

O. Zwikael and J. R. Smyrk, *Project Management*,
https://doi.org/10.1007/978-3-030-03174-9_4

A Project Governance Model (PGM) has two parts. The first (taking the form of an organisational model) identifies all the roles that will be recognised in the project and shows how those roles relate to each other. The second is a set of supporting role definitions.

The range of feasible management models that could be proposed for a project is very wide indeed. Casual observation of the way that typical business projects are organised suggests that many of the organisational structures adopted in practice are largely ad hoc with poorly defined roles leading to gaps/overlaps of responsibility, conflict amongst participants, inadequate processes and unreliable decisions. A governance model should be not only theoretically sound, but also have predictable behaviours to support the conduct of the project. It is appropriate therefore, to constrain the optional governance arrangements that might be considered for a project. The principles discussed below constrain the generic governance model for projects in three ways: by setting its general shape and features, by identifying the particular elements that are used in role definitions and finally by providing generic definitions for specific roles.

Because of the way that a project evolves, it is to be expected that a project governance model will also evolve in two ways: the overall "shape" of the model (reflected in a diagram showing how the elements of the project organisation fit together) and in the definitions attached to various roles.

It should be noted that if we locate projects on some spectrum of size/complexity, as we move toward the small end of the scale, the need for a formal project governance model fades (and the need for these principles also evaporates).

4.1.2 Principles of Project Governance

The approach to project governance is based on the following eight principles:

1. **Avoidance of conflict of interest**.
    - Description: A conflict of interest arises when someone could benefit (even indirectly) by acting against an assigned responsibility. It should be noted that a conflict of interest exists even if someone does not act improperly. When assigning project roles, conflicts of interest must be avoided.
    - Example: An external supplier is disqualified from participation in many of the decisions that must be made during the course of the project (such as scope change) and therefore cannot be a member of the steering committee. (This does not, of course, prevent a supplier from filling other roles as membership of a reference group, project team or project control group.)
    - Foundation: Avoiding conflict of interest is a common general principle adopted in many areas of commerce and law.
    - Rationale: Someone facing a conflict of interest may make decisions that do not favour the project. The principle reduces the risks associated with such an event.

2. **Project governance models are separate to organisational governance structures.**
   - Description: In general, a project's governance structure and the standing structure of an organisation have to be kept separate because they address distinct roles and responsibilities. It is undesirable (and unnecessary) to show project governance models as extensions to a standing organisational structure chart. The project owner and other members of the steering committee provide the most significant organisational connections required by a project governance model (refer to Figs. 4.4 and 4.5).
   - Foundation: Project-related accountabilities and responsibilities lie with those stakeholders who have been commissioned to fill specific roles in the project governance model. All these roles eventually (directly or indirectly) report to the steering committee whose members are appointed by the funding organisation(s). The brief for steering committee members, at the time of their appointment would normally outline the way in which they will report back into their home organisation. In some cases, similar external reporting arrangements will also be put in place for certain reference groups and advisers.
   - Rationale: A project governance model is intended to show how *the roles within a project* are arranged rather than how the *projects within an organisation* are arranged.
3. **A project owner is to be held accountable by the funder for target return.**
   - Description: a project owner is appointed by, and held accountable by, the funder for the achievement of the project's target return. A project can have only one project owner. A funder may elect to fill the role of project owner him/herself.
   - Example: Consider a major international project to reduce the incidence of child abuse. The various participating countries could have consistent (but distinct) target outcomes, for which each would make someone senior accountable. It would be appropriate to appoint all these country representatives to a steering committee. The appointment of the project owner for the project would be ratified by all participating countries.
   - Foundation: The principle arises from the ITO model, coupled with the principle of separation to emerge from principal-agent theory (Eisenhardt 1989). An implication of the principle is that a project manager cannot be held accountable for target outcomes (refer to principle 6 below).
   - Rationale: By making someone accountable for target returns we ensure that the interests of the funder drive the conduct of the project.
4. **A "purchaser" is to be separated from a "provider"**:
   - Description: Box 2.6 alluded to a project as a transaction between a supplier and a client. More formally, in any transaction, two separate entities should fill the (distinct) roles of *purchaser* and *provider*. In the case of a project, the funder (and hence the project owner), is the *purchaser* who is "buying" outputs

from a project manager (as *provider*). Accordingly, a project manager may be contracted from outside the funding organisation.
- Example: In a project to reduce travel time between the CBD of a city and suburbs on the other side of a river, a bridge can be viewed as the subject of a transaction in which the bridge is being "purchased" by the funder from a project manager (as "provider").
- Foundation: The principle (known formally as "the separation of purchaser from provider") is used widely, especially in projects seeking to reform government services for which there is no true market (such as public health).
- Rationale: The principle encourages clarity and transparency of the processes surrounding a transaction by separating the accountabilities of a "provider" and a "purchaser". The resulting relationship between the project manager and the project owner serves as the nucleus for the entire governance framework.

5. **An "agent" is to be identified and separated from the role of "principal"**:

   - Description: The funder of a project may (in the role of a principal) appoint someone else to look after his/her interests. Appointment of an agent is purely at the discretion of the principal. Because a project owner (as the funder's agent) holds an accountability for target outcomes, he/she must be appointed from within the funding organisation. If an agent is not appointed, then, effectively, the funder is self-appointing as project owner.
   - Example: Because of heavy time demands, the CEO of a company may appoint the national purchasing manager as the project owner of a project to reengineer the company's procurement processes. This appointment creates a principal-agent relationship between the two of them.
   - Foundation: The principle was first introduced by Eisenhardt (1989). Turner and Muller (2004) used the principal-agent theory to analyse this relationship in the project environment.
   - Rationale: Funders are usually quite senior in their organisations. As well as the demands on their time arising from their executive management roles, they also often face large portfolios of projects. This situation prevents them from playing day-to-day role in each project—and so they are forced to appoint agents to act on their behalf.

6. **A project manager is to be held accountable by the owner for delivery of project outputs**:

   - Description: a project manager is notionally engaged by, and held accountable by, the project owner for the production, delivery and implementation of the project's outputs.
   - Foundation: The principle arises from the existing practices of the project management profession (e.g. PMI 2017).
   - Rationale: By making someone accountable for outputs, the project owner can exercise appropriate control and influence over the direction of the work of the project.

7. **The role of each participant in the project governance model is to be formally defined and acknowledged as a condition of appointment**:

    - Description: Every significant role recognised in the project governance model is to be specified in some formal instrument (such as a charter, role description or terms of reference) which incumbents are required to acknowledge and accept.
    - Example: All members of the steering committee would be expected to accept a (fairly generic) charter which covers (amongst other things): the purpose and role of the committee, the scope of involvement in the project, and a confirmation that members will work towards the successful achievement of the projects baseline documents.
    - Foundation: The principle is grounded in organisational theory (Daft 2007) and human resource management literature.
    - Rationale: Formalising roles becomes, in effect, a mechanism for managing the risks associated with organisational overlaps and gaps. It also reduces the risks associated with misunderstanding of roles.

8. **The release of a resource to a project by a third party should be covered by a formal agreement**:

    - Description: Where the appointment of a person to fill a project role, requires the approval of another entity, then it may be appropriate to have the arrangement formalised with an appropriate instrument. If the approving organisation is a separate legal or commercial entity, then the instrument may take the form of a contract or commercial agreement. If the approving organisation is not a separate legal or commercial entity, then the instrument would take the form of a memorandum of understanding (MoU). An MoU lays out the terms of an agreement, but without the legal sanctions that underpin a commercial contract.
    - Example: If the project to reengineer the company's procurement processes required the release of an experienced purchasing manager from routine day-to-day work to participate in the design of the new versions of those processes, it would be appropriate for the project manager and the project owner to draft an MoU governing the purchasing manager's deployment to the project. By way of contrast, the engagement of a firm of process improvement specialists would involve *a commercial contract*.
    - Foundation: The principle is enshrined in long-standing business practice and commercial law (e.g. Kerzner 2017).
    - Rationale: Formal agreements ensure that everyone involved in the arrangement is clear on what is being offered, the conditions under which it is offered and what processes are to be followed in the event that conditions of the agreement are violated.

4.1.3 Project Governance and the Funder

A funder is an entity with discretionary authority to spend money on the project and is to be distinguished from *a source of funds*. For example, in a project to expand an online retail business, if the owner of the business arranges a loan from a bank, it is the business owner who becomes the project's funder—not the bank.

The role of funder does not appear explicitly in a project governance model (this is why there is no role definition for a funder). The entity who serves as a project funder may, nevertheless, self-appoint to the steering committee. In this situation, he/she is now recognised in a project governance model—but only as a member of the steering committee. The funder is an investor who is making a budget available to the project today—in the expectation of realising a flow of target outcomes tomorrow. The governance model adopted for a project seeks to organise all participants in a way that supports a successful investment. As discussed in the principles of project governance above, the funder frequently appoints a project owner to look after the project day to day. The project owner is, in turn, supported by a steering committee whose members are expected to always act in the funder's interests.

Because the project owner, therefore, speaks on behalf of the funder, the two will maintain close regular contact throughout the exercise, so that the funder is kept informed of the project's status. This close liaison will also allow the funder to endorse/confirm decisions that have been made by the project owner and provide input as necessary. It is through the steering committee in general and the project owner in particular that the funder remains as the dominant stakeholder in a project.

A project may have more than one funder. This will occur when it is not appropriate for one entity to bear the investment burden alone. The degree of concordance between the objectives of co-funders will shape the project's governance model. At one extreme, all funders may share identical expectations as they relate to objectives and outcomes. Such a project could be approached as if there were only one funder. At the other extreme, funders may have conflicting expectations, incompatible objectives and mutually exclusive target outcomes. Joint funding of such an exercise would make little sense, regardless of how it was governed. Between these two situations are ventures in which the prospective funders find their expectations are compatible, their objectives congruent and their target outcomes distinct but consistent. Multiple funders can have inconsistent views which might lead to contradictory advice concerning the direction of the project. The steering committee is responsible for resolving such conflicts.

4.1.4 The Involvement of Key Players in a Project

Involvement of key players in a project can take any of four forms—stakeholding, role, responsibility and accountability, as shown in Fig. 4.1.

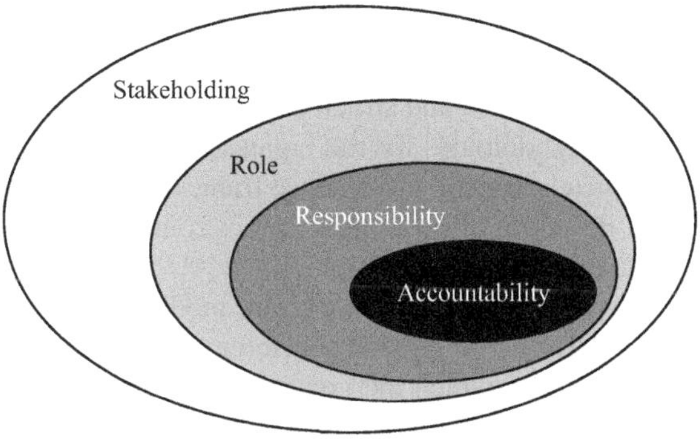

Fig. 4.1 The relationships between stakeholding, role, responsibility, and accountability

- **Stakeholding**. An entity has a stakeholding in a project when either the project has a potential impact on that entity or the entity has a potential impact on the project. All those who are commissioned to fill a role in the project are stakeholders by virtue of the fact that their appointment results in their having an interest in the venture.
- **Role**. A role for an entity arises when a project-related activity requires the involvement of that entity for its completion. For example, the utilisation of a project's outputs requires the involvement of project customers and so project customers fill a role in a project.
- **Responsibility**. A responsibility arises when a role is based on a formal agreement by the relevant entity to participate in certain activities. For example, a project team member is responsible for ensuring that all of the work of a particular project activity is appropriately executed, and so we would expect the arrangement for the activity requirements to be reflected in some sort of agreed brief between the project manager and the relevant team member.
- **Accountability**. An accountability arises when a responsibility is subject to agreed rewards and/or penalties. An accountability must also be accompanied by one or more *authorities* (powers to take specific actions or make particular decisions). Without authorities, accountabilities collapse into responsibilities. Similarly, the existence of rewards and sanctions converts a responsibility into an accountability. For example, if the project manager is held accountable for the delivery of all project outputs (fit-for-purpose, on time and within budget) then he/she must also be granted the authority to deploy the resources made available for the conduct of the exercise. Accountabilities should be confirmed in a formally agreed instrument of some kind–such as a memorandum of understanding or a contract (see principle 7 above). It is important to note that, in general, a project customer has no responsibility for (and cannot be held accountable for) utilising the projects outputs.

Of these four forms of involvement, only those stakeholders who have a role, responsibility or accountability in the project are included in the project governance model.

4.1.5 The Structure of the Project Governance Model

Numerous models and structures have been proposed at different times and in different methodologies for the organisation of projects. Here, we attempt to provide a completely general governance framework based on the principles defined above. Most, if not all, of the approaches in common use can be mapped to this model as special cases.

The proposed project governance model is made up of four *divisions*, as presented in Fig. 4.1: steering, delivery, reference and assurance. These divisions provide the foundations on which the governance arrangements for each project are built (see Sect. 4.1.6 below). Two of these (steering and delivery) must appear in the project governance model of every project. The reference and assurance divisions are not always necessary. The governance models differ therefore among projects at three levels: in the need for the reference and assurance divisions, in the structure of the reference division (when it exists) and in the membership of the particular classes of role that are created in each division. It should also be noted that the governance model for a particular initiative will evolve over the life of the project in these same three levels. In other words, changes over time will be noted both in the number of classes represented within each division and in the membership of ongoing classes. For example, it is conceivable that a reference group may exist only for a single working session.

1. **Steering**. This division is concerned with leading the project so that it remains consistent with its baseline documents. There are two classes of role in this division—the project owner and the steering committee (which the project owner normally chairs). The steering division is recognised in all projects.
2. **Delivery**. In this division, all those responsible for producing, delivering and implementing the project's outputs are located. There are up to four classes of role here: the project manager, project administrator, project control group and the project team. The delivery division is recognised in all projects.
3. **Reference**. In this division, all those who have been commissioned to provide specialised input to either the team or the steering committee are located. Not all projects require these roles. Reference division appointments are made for either of two reasons:

    - With the specific intention of engaging particular spontaneous stakeholders in the project (see Chap. 5).
    - To obtain input from someone who cannot meaningfully be employed as a project team member—for example, the services of a lawyer.

    Reference division appointments may or may not be remunerated.
4. **Assurance**. Not all projects require assurance roles. In this division, those responsible for independently monitoring the conduct of the project on behalf of the steering committee are located. The two most common assurance roles are those of project assurance counsellor and probity counsellor. It should be noted that

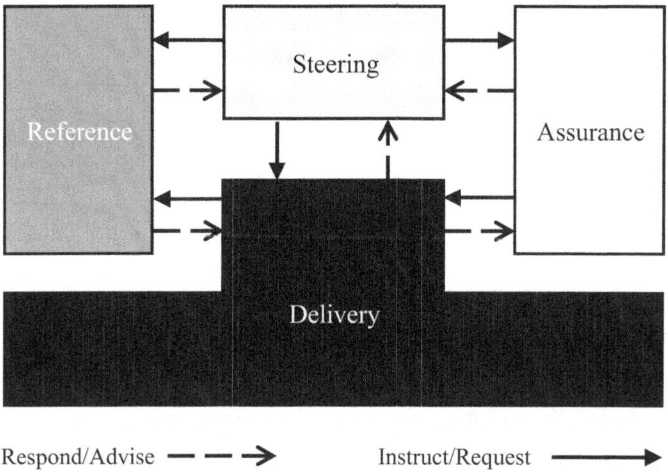

Fig. 4.2 The four divisions of the generic project governance model

both these are *assurance* not *audit* roles. This allows counsellors to advise and work with project managers as they see fit.

Project assurance roles are strictly above-the-line and are to be distinguished from those related to *output* quality control/assurance. Such quality management roles are strictly below-the-line, and hence appear in either the delivery or reference divisions.

The relationships between these divisions (as reflected in the arrows appearing in Fig. 4.2) bear further discussion because they determine how all entities in a specific instance of a model eventually interact. Note that, the arrangement of relationship arrows is not symmetric.

The steering committee acts on behalf of the funder. Because it approves the detailed design of the project governance model, all those who play a part in the exercise are eventually accountable to the steering committee for successfully discharging their responsibilities.

Each entity in the reference division "reports" (in the line sense) to either the steering committee or the project manager. While the project manager, for example may *brief* a reference group at its regular meetings, he/she is never subordinate to, or instructed by, such a group (otherwise it would become a "competitor" to the steering committee).

Although assurance counsellors (in those projects where they are appointed) can *instruct* the project manager to provide required management information about the project, in practice they will be working closely and cooperatively with the project manager and so any information for quality assurance purposes is more likely to be *requested*.

The project team is closely linked to the steering committee (through the project manager) by formal reporting arrangements including regular review meetings.

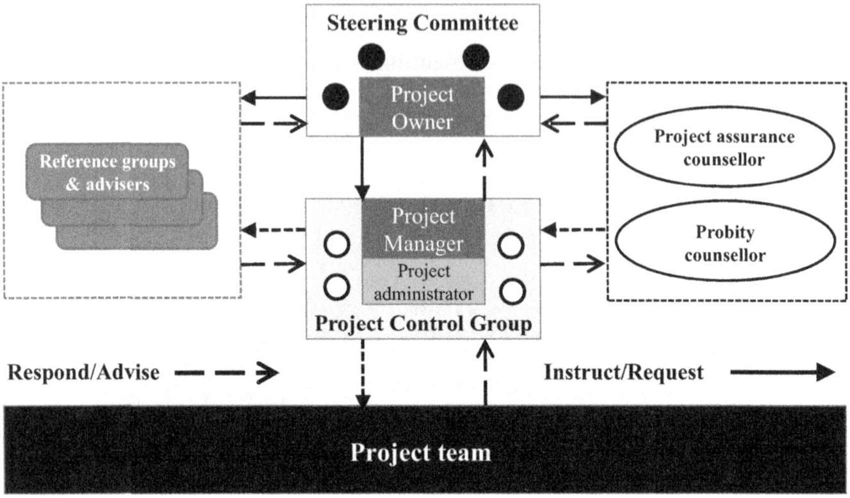

Fig. 4.3 The detailed project governance model showing the classes of role within each division

4.1.6 Designing a Project Governance Model

Nine project roles are recognised within the four divisions of the governance model, as shown in Fig. 4.3.

A project governance model is *defined* when:

- The roles peculiar to a project are identified.
- The relationships amongst those roles are identified.
- A definition is assembled for each role.

The first two of these requirements is most readily displayed as a project governance model diagram. This will take the form of a specific instance of Fig. 4.3. The structure of a role definition is provided in Table 4.1 and specific definitions of all core project roles are presented in Appendix B. A description of the nine governance roles is provided below:

1. **The steering committee** is made up of powerful supporters of the project (including the project owner). It is the ultimate authority in the project governance model, approving baseline documents and directing project execution so that those baseline documents can be realised. The steering committee supports the project owner in discharging his/her accountabilities. All projects have a steering committee—even if only a "committee of one" (the project owner).
2. **The project owner** directs the project—as the funder's agent.
3. **The project control group** supports the project manager in discharging his/her accountabilities. The project manager normally chairs the project control group which is made up of senior people from those organisations who provide resources to the project. Through the project manager, the project control group

Table 4.1 A generic template of a role definition for a project governance model

Attribute	Description
Name of project governance model role	The label attached to this role in the project governance model diagram?
Division in which this role is located	Identify in which of the (four) divisions of the generic PGM diagram this role appears
Objective of the role	As a "one-liner", state why this role is important to the project
Accountability	This entry is required only for the project owner and project manager
Outputs from role	List all the substantive outputs that will emerge from this role
Core activities and frequencies	What activities does this role entail? How frequently will that activity take place?
Membership	How many people will fill this role? How will they be selected? If known who are they? If it is a group—who will lead it?
Communications programme	What special forms of communication (outside the project's formal reporting framework) is required for whoever is assigned to this role?
To whom does this role report	Who instructs the role to do work?
Review of role (who/when)	Who will review the effectiveness of the role? When will this be done?
Dates of creation and termination	When does this role come into existence? When does it cease to exist?

manages project execution. In a small project the project control group will have just one member (the project manager). It is worth noting that the diagrammatic representation of the project control group "mirrors" that of the steering committee.

4. **The project manager** leads and guides the project team day to day (Unger-Aviram et al. 2013). All projects will have a (single) project manager.
5. **The project administrator** supports the project by overseeing the project's above-the-line work. Project administrators are only required when the project's above-the-line work cannot be adequately resourced by the project manager or selected team members. Large projects may have more than one project administrator.
6. **The project team** is made up of all those who resource the project under the direction of the project manager. Some resourcing is also done through reference appointments—but the work involved in that case is not managed directly by the project manager. Typically, the team would be made up of appointed staff, suppliers, consultants and contractors. The structure adopted for a team will depend on the peculiarities of the project and could be based on skills, functional areas, outputs, contracts or type of work. All projects have a project team.

7. **Reference groups and advisers** may or may not be required in particular projects. There are two situations which give rise to the creation of a reference group (or the appointment of advisers):

   A. It has been decided that, in order to engage particular spontaneous stakeholders (see Sect. 5.1.2), they will be *invited* to join a reference group—or serve as advisers. The primary motive for issuing such an invitation is to establish a positive relationship between the stakeholder and the project's key players.
   B. It has been decided that someone is to be engaged, in order to gain access to them as a project resource regardless of whether or not that person is a spontaneous stakeholder in the project. If, for any reason, it appears inappropriate to have that person join the project team, then they would be appointed to reference group—or as an adviser. In general, if the *work* to be undertaken by the appointee is subject to oversight by the project manager, then that role is best located within the project team. By way of contrast, take the example of a lawyer engaged to review a supplier contract. Normally, the project manager would have no interest in *how* this work was undertaken—only in the *output* from that work. Such a role would be best located in the reference division.

   In both cases, an appointment to a reference group (or as an adviser) is governed by a role definition—as discussed above in principle 7.
   A project could have numerous reference groups and advisers—some of which may have extremely short tenures. They may be given titles such as "reference group", "task force" or "working party", while those that involve individuals will usually be identified as "advisers".
8. **A project assurance counsellor** is appointed to ensure that the project is conducted in accordance with the project management framework adopted by the funder (which could also recognise any peculiar internal project management frameworks employed by subcontractors).
9. **A probity counsellor** is appointed to ensure that the commercial dealings between project participants and the outside world are being managed in accordance with the commercial guidelines of the organisation.

Two questions must now be answered if a project is to be successfully accommodated within an organisation's standing structure:

1. How is the project to be linked into the organisational structure?
2. How is competition for resources (between projects and operations) to be resolved?

Both of these questions are addressed in the following two sections.

4.1.7 Project Governance in an Organisational Context

Appropriate linkages must be established between the (transient) structure adopted for the project and the (standing) structure adopted for routine business purposes by the funding organisation. Such linkages represent (and allow for) flows of:

- **Information**. From the project to the standing structure—such as advice about the current status of the project budget: "The approved budget is expected to be exceeded by 20%".
- **Instruction**. From the standing structure to the project—such as declarations of policy: "Only locally produced structural steel is to be considered for purchase".
- **Authority**. From the standing structure to the project—in the form of guidelines for decision-making and expenditure: "The project owner is authorised to approve supply contracts of up to $10,000".

By way of contrast, the linkages one finds within an organisation's standing structure also indicate other forms of relationship between positions, such as performance assessment, reward and career progression.

While it is common to see charts that attempt to locate projects in their organisational context, the value of such diagrams is far from clear. Because of the linkages involved, such figures present some very awkward topological problems for which there are no simple solutions. The creation of (usually highly artificial) arrangements under which project steering committees report to organisational committees, for example, may make the diagramming easier, but their contribution to better management is questionable.

How then does an organisation connect to its projects? Two distinct funding situations require different approaches. If there is only one funder, the project can be treated as a virtual business unit in which the project owner—as "head of" (or leader) reports into the organisation's standing structure via the operational/functional unit where the funder is located. If there are multiple funders, such an arrangement may still be employed—in which case, all the funders appoint one of their number as their (common) agent (see Fig. 4.4). If, however, such an agency model is not acceptable, then a number of linkages between a project and the funding organisations are created through members of the steering committee in general and the project owner in particular, as suggested by Fig. 4.5. Information presented to the steering committee subsequently becomes available to the funders via these channels. At the same time, instructions and authorities (especially those concerning deployment of resources) from the funders are relayed to the project via members of the steering committee.

In addition to these formal linkages, there will normally be numerous informal paths along which information flows out of the project into participating organisations. (Each representative appointed to specific roles will obviously provide such a line of communication). In most situations, it is not necessary (or even useful) to show such (ad hoc) linkages between projects and their participating organisations.

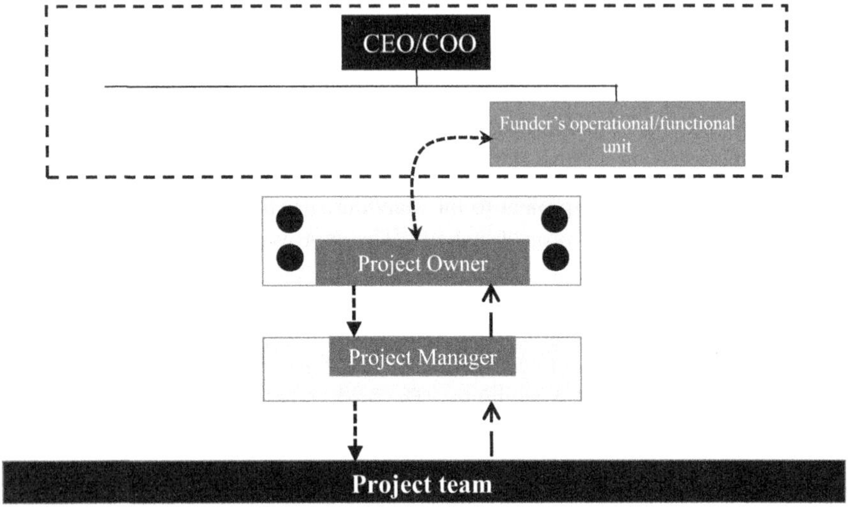

Fig. 4.4 An example of linkage between a governance model and a single funding organisation

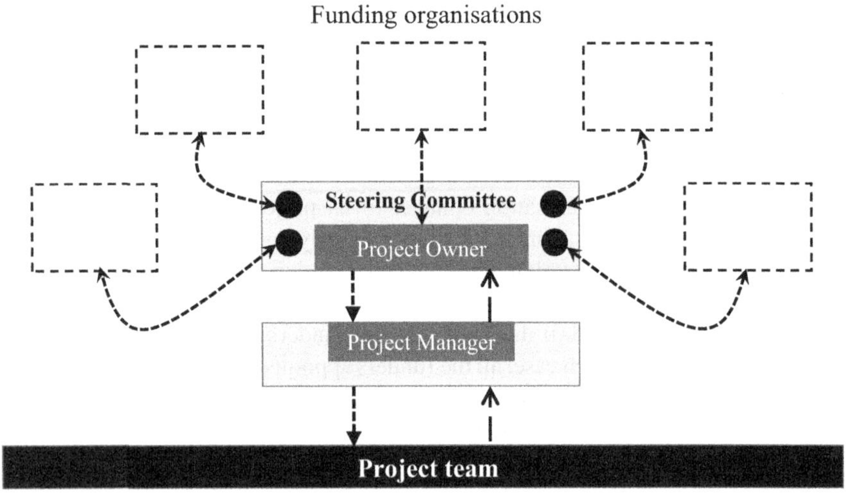

Fig. 4.5 An example of linkage between a governance model and multiple funding organisations

4.1.8 Project Governance Resourcing Issues

As well as purchased-in resources, most projects make demands on staff of the funding organisation to resource both below-the-line and above-the-linework. In the case of below-the-line work, it is critical that knowledge of, and experience with, downstream processes be made available to the project. This knowledge and experience can only come from those staff who participate in "business-as-usual" processes. A consequence of this is that, inevitably, projects compete for the same resources that are required to maintain routine business operations. Because, in general, operations must take precedence over projects, any unavoidable resource contention will result in delays to the project's schedule.

There are three broad arrangements under which operational/functional staff can be made available to a project team:

- A permanent secondment in which the staff member is assigned full time for the duration of the project.
- A short-term secondment in which the staff member is assigned full time for a designated period of time.
- A part-time appointment in which the staff member alternates between project and operational duties.

Part-time appointments automatically involve dual lines of reporting in which the staff member is subject to the parallel demands of a line manager and a project manager. Even in the case of secondments, conflicting demands can arise, especially if the engagement is for a relatively short period of time. This is how a matrix structure emerges. Note that, a matrix structure is not so much a model of management, as it is simply a phenomenon arising from multiple lines of reporting. Matrix structures become problematic when they give rise to resource contention. This problem is exacerbated by fluctuating loads and the associated swings in demands for staff time from both the project and the organisation's routine operations. If we use a week as a representative resource-planning horizon, then clearly the total weekly workload on a part timer from both the project and the organisation will rarely equal his/her availability over that week. When the load exceeds availability, then the employee faces potentially conflicting and irreconcilable demands.

No matter how the team is organised, dual reporting lines bring with them resource contention problems (as reflected in Fig. 4.6)—and so the focus is now on how to manage the resource issues that arise from a matrix structure. While the diagram uses the case of a part timer, it is completely general in its interpretation—a seconded full timer can also be faced with a similar problem.

Amongst the strategies to deal with this problem, three are noteworthy. The first concerns the formal acknowledgement of roles identified in the project governance model. According to principle 8 (discussed in Sect. 4.1.2) all internal appointments should be formalised in a Memorandum of Understanding (MoU) which should establish very clearly key elements of the appointment including:

- The role to be filled by the appointee.
- The period of the engagement.
- The intensity of the appointment (full time, part time).
- A nominal level of availability for the project.
- Guidelines on how the MoU should be managed.
- A strategy for redeploying the staff member at the end of the assignment.

The document would also note the emergence of a matrix structure and acknowledge the possibility of resource contention, as well as outlining mechanisms for

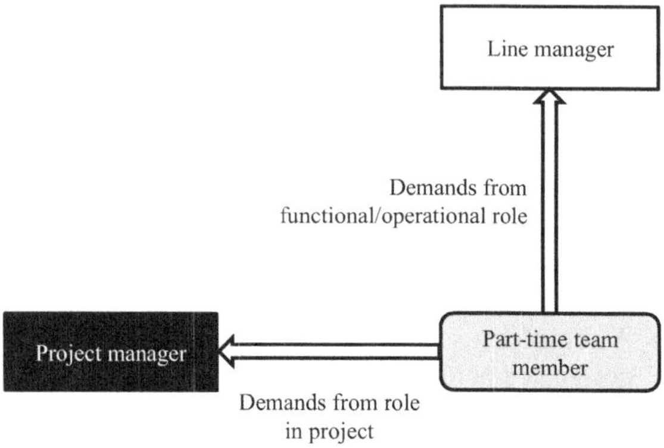

Fig. 4.6 A matrix structure emerging from part-time project appointment—causing resource contention

dealing with such situations. The MoU would be signed off by the staff member, the line manager and the project manager.

The second approach concerns the relationship between the project manager and the line manager. Resolution of resource contention problems will involve both players and so the soundness of their relationship will be a significant factor in the effectiveness of their problem-solving actions.

The third approach involves a critical role for the steering committee. Participating in the resolution of resource contention problems is a key function of the steering committee and should be reflected in its charter and structure.

4.1.9 Projects and Contractors

We now turn our attention briefly to an issue that causes considerable confusion in the discussion about projects in general and project governance in particular. This relates to the question "Does the term 'project' mean the same thing to a funder as it does to any contractors who might be involved?" At best, the question is left unanswered—at worst it is answered in the affirmative.

Take for example, a home improvement project involving: a new wing (with additional rooms), new furniture and landscaping. The homeowner (who is the project funder) has selected and separately contracted:

- A renovations company to do the new wing.
- An interior design firm to do the interiors and supply furniture/fittings.
- A landscaper to do the garden.

Because their respective contracts are so significant, the renovations company, the interior designer and the landscaper each see fulfilment of their contract (understandably) as a "project". In other words, *the scope of the contract completely determines the scope of a contractor's "project"*.

4.1.10 *Project Governance and Above-the-Line Resourcing*

The load represented by the project's above-the-line work must be resourced. At one extreme, in small projects, this falls to the project manager. As a project increases in size, it may be feasible to assign some of this work to team members. For example, the maintenance of the stakeholders, risk and issues registers (discussed in Chaps. 5 and 6) could possibly be delegated to particular team members—who then become the stakeholder manager, risk manager and issues manager, respectively. Unless the team just happens to include suitably qualified members, it may not be possible to farm out work in this way. Eventually, it becomes necessary in larger projects to appoint a project administrator to look after all of a project's above-the-line work. In fact, very large projects may require the establishment of an entire project administration sub-team.

4.1.11 *The Operation of the Project Governance Model*

Not only will the shape of a project governance model be peculiar to each project, it may also evolve over the life of the project. In what follows we discuss the model shown in Fig. 4.3 in its entirety, although all of the components will exist only in large projects. The focus of this immediate discussion is confined to project execution (which is where the bulk of governance processes are carried out).

The *information* flows (shown as dashed arrows in Fig. 4.3) may be either requested or volunteered. It is important to note the directions of the "instruct/request" arrows. In particular, under no circumstances do reference groups or advisers issue instructions to anyone else.

While it is important to maintain good working relationships throughout the project governance model, three are critical:

- Between the project owner and other members of the steering committee.
- Between a project assurance counsellor and project manager.
- Between project owner and project manager (see Box 4.1).

All three of these relationships should be the subject of a "no-surprises" style of management.

Through the project owner, the steering committee is, in effect, the project manager's "client". Likewise, the project manager is, in effect, the steering committee's "supplier" (of project outputs). In the event of differences of opinion amongst its members, the steering committee must reach a consensus on the instructions and guidance it issues to the project manager.

Reference groups and advisers should be tightly briefed. This means that: their role definitions should be absolutely explicit, their memberships held to a minimum, their tenure short and their reporting lines clear. The upshot of this is that a large number of small, short-term, narrow focus reference groups emerge in the project governance model (instead of a small number of large long-term groups with wide terms of reference).

During execution, regular periodic project status reports will be tabled by the project manager for consideration by the steering committee. The steering committee will, in turn, decide on appropriate courses of action to be taken by the project manager. The steering committee may request input to its oversight of the project from reference groups—as well as from the two counsellors. References groups and counsellors may also initiate flows of information to the steering committee.

More is said about periodic project status reviews in Chap. 11, but a brief note on these helps understand how the project governance model operates. Discussion about the project's progress takes place in a variety of forums, including:

- The steering committee
- The project control group
- The project team
- Sub-teams
- Subcontractors, consultants and suppliers.

Meetings of the steering committee will typically take place every month. These meetings are relatively formal—in that discussion is guided by an agenda, with decisions and agreed actions being minuted. Various reports are tabled and presented by the project manager for consideration by the steering committee. Meetings of the project control group and project team are normally held every week and although often somewhat less formal, they are, nevertheless, guided by an agenda (similar in coverage to that of the steering committee) and also minuted. The other forums and their conduct tend to be peculiar to each project, as well as the intervals between meetings proposed above.

From the literature

Box 4.1: Control and trust strategies for the relationships between the project owner and project manager

Zwikael and Smyrk (2015) conducted an empirical study into the relationships between the project owner and project manager. They compared the effectiveness of two management approaches to support the role of the project owner—control and trust. Control is a formal mechanism which is associated with legal documents, such as a contract, which enables the control of processes surrounding resource allocation and decision-making. Trust is often expressed in terms of confidence in another's intentions, or faith in a partner's moral integrity. Trust is associated with mutual understanding, unrestricted learning, and inter-organisational knowledge sharing and can substitute a more formal control mechanism by avoiding several types of transaction costs, while still ensuring the other party meets their obligation.

Results suggest that trust of the project owner in the project manager is more effective in a turbulent environment, whereas more control by the project owner of the project management process is a superior management approach in a more stable project setting.

4.1.12 Managing the Project Governance Model

The project governance model is assembled progressively over the early phases of a project. During initiation, the project champion (in consultation with the funder and project-manager-designate) designs a broad model by identifying the roles that need to be filled and deciding on the most appropriate assembly of elements corresponding to those roles. While the identity of candidates for certain roles might be known at this time, it is not possible to complete the first version of the model until the planning phase. There will be occasions where a form of governance is required during the planning of a project (as might happen on very large projects). In those cases, the associated preparatory stakeholder engagement and establishment of the project governance model will all be carried out as an additional set up activity early in planning.

The project governance model (together with supporting role definitions) is formally approved by the funder when accepting the business case, then any subsequent changes are approved later by the steering committee and, where necessary, also ratified by the funder.

Progressively throughout initiation, the project governance model is assembled collectively by the champion, the funder, as well as the project owner, project-manager-designate and the project assurance counsellor (if appointed). This is done in two steps:

- Designing the governance model, the output from which is a PGM diagram which identifies all the roles that are to be recognised.
- Defining each role—guided by a structure like the one suggested in Table 4.1. Many of these definitions (for example, that of the project manager) will be generic to an organisation. Others will require "hand-crafting"—especially those for all roles recognised in the reference division.

4.1.13 Project Governance and Professional Development

Casual observation about the effectiveness of appointees to a project governance model suggests a wide spectrum of proficiency. This is particularly the case with those who fill a role outside the project team. Steering committee members, for example, are chosen because they are powerful supporters rather than because they have background in the project management discipline. At the same time, they need a good understanding of their role and the processes that surround it. It is, therefore, desirable that programmes of skill enhancement be offered to those participants who are to fill roles in both the reference and steering divisions.

4.1.14 The Project Management Office (PMO)

While all projects demand the approval and sponsorship of executive management, some also merit the support of additional specialists with above-the-line skills and resources. In recent years, many organisations have established a Project Management Office (PMO) to assist those involved in projects with facilities such as programmes of professional development, templates, guides and computer tools. Although the PMO is not part of the governance of any single project, it is an important supportive entity to the projects executed in a particular business unit. Distinct forms of charter for a PMO can be assembled by selecting from the following list of candidate functions:

- Resourcing: the provision of specialist above-the-line skills to projects—in particular: planning, management, administration and assurance.
- Capacity development: Mentoring, professional development and training programmes for staff and (in larger organisations) sponsorship of communities of practice.
- Methodological guidance: assembly/adoption and implementation of project management frameworks.
- Standards: Ensuring adherence to local practice and performance evaluation.

Other responsibilities appear in practice from time to time, including portfolio management, project funding and ownership. Assignment of roles of that nature to a PMO would be difficult to reconcile with the governance principles discussed above. Various definitions and roles of the PMO are summarised in Box 4.2.

From the literature

Box 4.2: The Project Management Office (PMO)
The Project Management Institute defines a project management office as follows: "an organisational structure that standardises the project-related governance processes and facilitates the sharing of resources, methodologies, tools and techniques" (PMI 2017, p. 48). Another definition for the project management office by Ward (2000) is: "A project management office is an organisational entity established to assist project managers, teams and vari-

ous management levels on strategic matters and functional entities throughout the organisation in implementing project management principles, practices, methodologies, tools and techniques." Aubry (2015, p. 20) notes, the PMO is "an organizational entity assigned a variety of roles or functions in executing the coordinated management of projects under its domain." Finally, Korhonen et al. (2016) describe the role of a PMO as responsible for coordinating multiple projects.

Project management offices are created to help in various tasks of project execution and to integrate project-related work inside the organisation. Dai and Wells (2004) make a distinction between a project office (for a single project) and project management office (for the management of multiple projects in an organisation or business unit). Hurt and Thomas (2009) argue "A PMO can take many forms, ranging from simply providing administrative support for projects to providing coaching (e.g. as a center of excellence, on project management practices, tools, etc.) to acting as a full-blown execution function whose mandate is to formally manage and deliver projects for the organization". Korhonen et al. (2016) described the role of a PMO as responsible for coordinating multiple projects. Koh and Crawford (2012) indicated that PMOs are "involved in business strategic planning, ensured projects were carried out in time, budget and scope, managed risks, and made projects reviews, coaching, issue handling, and corporate process improvement after the project."

Letavec (2006) distinguish between three PMO functions: PMO as: (1) a consulting organisation; (2) knowledge organisation and; (3) standards organisation. Rad and Levin (2002) introduce project-focused and enterprise-oriented functions of a PMO. The project-focused functions include: consult, mentor, and augment. The enterprise-oriented functions include: promote, archive, practice and train. Hill (2008) introduces five distinctive PMO functions and their sub-functions. First, practice management, including the sub-functions of project management methodology, project tools, standards and metrics and project knowledge management. Second, infrastructure management, including the sub-functions: project governance, assessment, organisation and structure, and facilities and equipment support. Third, resource integration, including the sub-functions of resource management, training and education, career development, and team development. Fourth, technical support, including sub-functions of mentoring, project planning, project auditing and project recovery. Fifth, business alignment, including the sub-functions: project portfolio management, customer relationship management, vendor/contractor relationship management and business performance management.

4.2 Programme Governance

We define a programme as a collection of projects that require coordination. The need for coordination arises because of various types of interactions between projects—including:

- Target outcomes which are similar, shared or in conflict.
- Downstream processes which are common, or which interact during their execution.
- Project customers who are common.
- Outputs which are common.
- Workplans which are interdependent.
- Common resources for which the projects compete.

If not consciously managed, such interactions can reduce the attractiveness to funders of projects as investment opportunities. Conscious management, on the other hand, may not only reduce the negative impacts of interaction, but actually exploit them to enhance a project's attractiveness.

Coordination takes the form of linkages between the project governance models of related projects. These linkages give rise to programmes. Accordingly, the programme environment can be viewed as an extension of the project environment. It is important to note that because a programme is simply a coordination framework for a collection of projects, it has no separate target outcomes or outputs.

4.2.1 The Conditions Under Which Projects Should Be Coordinated

Before discussing the issue of how a collection of projects might be linked, we need to examine the opposite question: "Under what conditions is it meaningful for a funder to divide a project into a set of (smaller) projects"? The answer involves careful adherence to the definition of a project. A project can be decomposed into smaller projects only if each component has a valid scoping statement (with a statement of objective, target outcomes, committed outputs and downstream processes). If a project can be partitioned in this way, then the funder has the option of either conducting it as one venture—or breaking out the component projects for separate treatment. Such a decision may have significant implications for success of the exercise. It is not always obvious when a project can be decomposed.

If an initiative has outputs, but cannot generate target outcomes on its own, then (if it is to be undertaken at all) it can only represent a component of a (larger) project—it cannot be a project in its own right. Now having said that, we quickly run into a problem of terminology. Consider a project to reduce the costs of access for heavy trucks to the general cargo area of a large seaport involving: a new bridge over an estuary, a widened road and modifications to a number of underpasses (to increase their clearance). In scenario "J", access costs will not change unless all three outputs are delivered, while in scenario "K" delivery of each output would be associated with incremental decreases in access costs. Under scenario "K" the initiative could, if desired, be broken into three projects (each supported by a separate business case). In fact, in that situation, it may even make sense to fund only one or two of the individual projects. Under scenario "J", however, the exercise cannot be broken up in that way and so a single business case must embrace all three outputs. Assume that, despite this, the business case for scenario "J" imposes no particular constraints on when each output must be delivered. It could well be that scarcity of funds or resources force the proponents into a strategy of progressive delivery over many years. If the underpasses are to be modified in year 1, the bridge erected in year 3 and the road widened in year 6 (with lengthy periods of inactivity between them), how are we to refer to the three individual exercises—and to the overall initiative of which they are a part? While the formality of the approach presented here requires that the overall initiative be viewed as a *project* made up of three *sub-projects*, we do acknowledge that such terminology sits awkwardly with more common parlance (in which it may well be described informally as a "*programme* of *three projects*").

An interdependency amongst projects will emerge in either of two situations:

1. **Imposed dependency**, where it is discovered that the worth (the value to the funder) of one project will be reduced because its execution is constrained by the execution of another project. This sort of relationship has the effect of increasing duration or cost (or both), relative to what would have happened without the dependency. For example, the duplication of a rail line between a mine and a port may be delayed by urgent flood mitigation work on a nearby river. It is clear that in order to prevent unnecessary and avoidable loss of worth (due to cost and time increases), both projects should be scheduled and monitored in a coordinated way.
2. **Elected dependency**, where it is discovered that the worth of one project could be increased by voluntarily coordinating its execution with the execution of another. Consider two (currently) unrelated projects: one to reduce groundwater pollution by connecting all houses in a suburb to a new sewage treatment plant and another to increase community access to a range of services by running a fibre-based high-speed broadband network to those same houses. It may be possible to significantly increase the worth of one or other of these projects by integrating their planning and management, so that costs and durations (or both) can be reduced. For example, the trench for the sewer pipework may be modified to accept a fibre optic cable.

Both forms of dependency can arise spontaneously (as might happen when the projects are undertaken by different funders) or intentionally (when a collection of related projects is purposely assembled by a single funder, perhaps to give effect to a strategy or policy). Consider strategy by an IT company to reduce its reliance on manufacturing by moving into services (such as IT outsourcing and consulting). To achieve its objective, the firm may well commit to a programme of related initia-

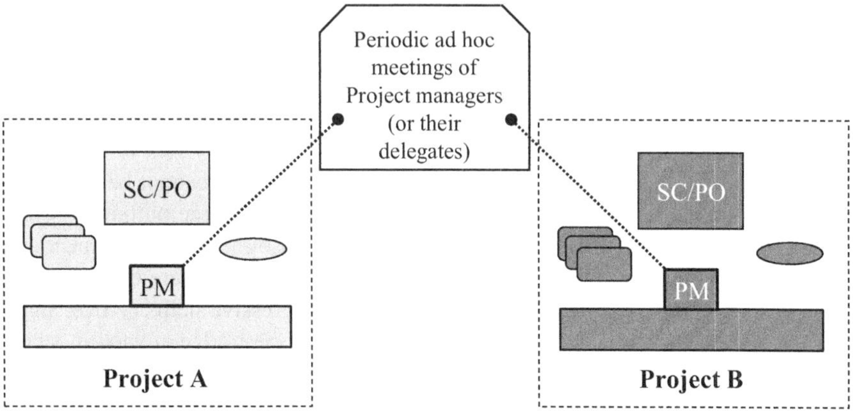

Fig. 4.7 A programme of *loosely coordinated* projects SC = Steering committee; PO = Project Owner; PM = Project manager

tives such as mergers and acquisitions, service development, staff development, skill acquisition, sales force redeployment and asset disposal.

Some of the items in this list may make sense as projects in their own right while others may, in fact, represent groups of related (and yet to be articulated) projects. Eventually as the component projects are defined, any dependencies amongst them will become apparent—as will any need for coordination.

4.2.2 Alternative Models for Coordinated Projects

Regardless of whether the relationships between projects are imposed or elected, three stylised forms of coordination are proposed–distinguished by the level of communication and harmonisation required. Projects are linked into programmes through their governance models. Note that the introduction of a programme does not automatically require the role of a programme manager, this depends on the type of coordination required as discussed in the three models below.

1. **Loosely coordinated projects**: Establish an informal arrangement whereby the project managers from each project maintain regular ad hoc contact to resolve coordination issues as they arise. This approach becomes unwieldy as the number of related projects grows (and, in fact, may prove impractical with more than two or three). An example of a governance model of a loosely coordinated programme is shown in Fig. 4.7 where the size of the shape representing the coordination forum has been exaggerated here for emphasis.
2. **Tightly coordinated projects**: Establish a (single) common coordination working party (which then sits within the reference division of each related project governance model) to consider planning options for each project and make rec-

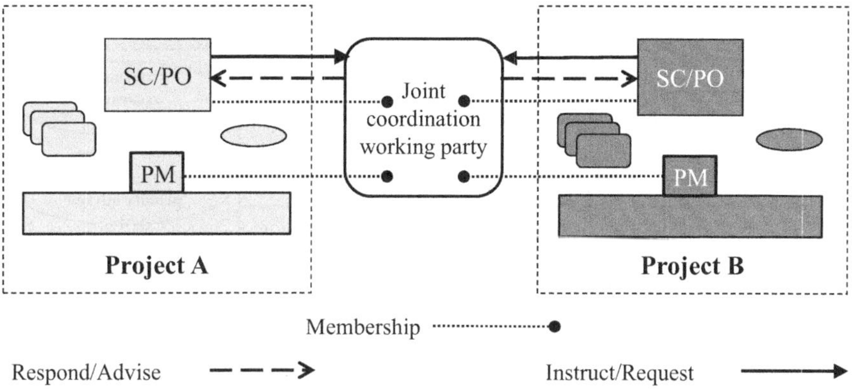

Fig. 4.8 A programme of *tightly coordinated* projects (with a joint coordination working party); PM = Project manager; SC = Steering committee; PO = Project owner

ommendations accordingly. An example of a governance model of a tightly coordinated programme is shown in Fig. 4.8 where:

- The Joint Coordination Working Party (JCWP) is common to both projects (and thus appears within the reference division of each project governance model).
- Again, the size of the shape representing the JCWP has been exaggerated for emphasis. In the project governance model for each project, it would look the same as all the other reference groups/advisers.
- As a reference group, the JCWP will have a formal role definition, which would include the formulation of a coordination strategy for the projects in the programme. In addition, the role definition would set an expectation that each project steering committee would tend to accept the recommendations of the JCWP.
- In view of the importance of the JCWP, it would normally include the project owners and managers from the projects being coordinated—who, between them, have the authority not only to formulate coordination strategy for the programme, but also to act on that strategy.
- Recommendations for actions to coordinate the respective projects are directed at the steering committees—who, in turn instruct the project managers accordingly. In any event, all recommendations must have the support of the respective steering committees.

3. **A programme of integrated projects**: When projects funded by the same entity require close coordination they can be placed under the control of one steering committee and one programme owner, one programme manager, with project managers as required. An example of a governance model of an integrated projects programme is shown in Fig. 4.9. Note that in such case, there is no need for project owners for each project, as the programme owner will be assigned

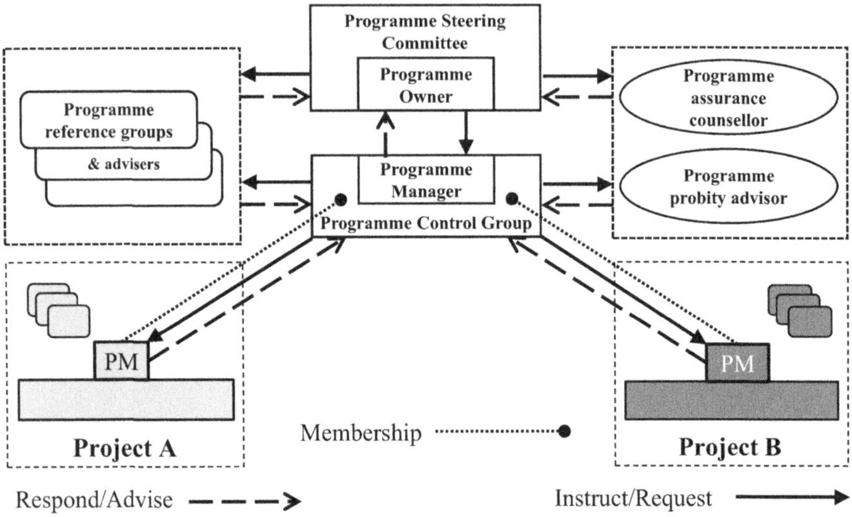

Fig. 4.9 A programme of integrated projects (with a programme manager); PM = Project manager

this accountability for the entire programme. Having said that, when assembling the steering committee, it would appear desirable to include those who might otherwise have served as project owners of the component projects.

It will be obvious that as we move from the first to the third option, the strength (and presumably the effectiveness) of coordination is increasing, however, so too is the severity of the constraints on each project. Clearly, the prospects for coordination are heavily influenced by any relationships amongst the funders. Disparate funding makes projects much more difficult to coordinate than funding from a common source. For this reason, the third option (Fig. 4.9) would be feasible only when there is a single funding entity. Thus, the third option would be open to the IT company mentioned above that has proposed a strategic move out of manufacturing into services. In some cases, this issue could be resolved by creating a special commercial entity (joint venture) to act as the funder.

Integrated projects are obviously run as (single) projects, except for the substitution (where appropriate) of the word "programme" for "project". So, for example in the governance model, there would be a programme owner, a programme manager (to whom would report "project managers").

The discussion of coordinated projects can be summarised thus:

- A programme is a collection of projects whose execution is coordinated.
- Projects are coordinated through their governance models.
- Projects may be linked into programmes for either of two reasons: they interfere with each other (because of an imposed dependency) or their worth can be increased by electing to coordinate their execution.

- Projects become candidates for linkage fortuitously or purposely (when they form part of a coherent strategy).
- Three models of coordination are consistent with the principles of governance proposed here: loose coordination (through cross-project representation), tight coordination (through harmonised management) and project integration.

The decision to coordinate a group of projects as a programme is taken very early in initiation. From that point on, the entire programme accords, in general, with the frameworks proposed here.

Summary

In this chapter, we have examined the engagement of key players and their organisation within a project's governance model. We also explored the concept of a programme and noted that programmes are handled in a fashion very similar to that suggested for projects. In the next chapter, we discuss the management of project and programme stakeholders, covering those included in the governance model as well as those outside it.

Chapter 5
Stakeholder Management

Abstract Each project has multiple internal and external stakeholders on whom the project can have a significant impact or who can impact its performance (either positively or negatively). The form of impact the second group can bring to bear may well be a response to the way they are affected by the project. For this reason, it is important stakeholders are managed carefully throughout the project's life. This chapter discusses six classes of spontaneous stakeholders and nine classes of commissioned stakeholders in a project. In addition, we discuss a five-step stakeholder management process and propose two practical tools to support this process: a stakeholder register and a stakeholder report.

5.1 Project Stakeholders

A stakeholder is defined as an entity who can impact the project and/or be impacted by the project. Stakeholders add to the management challenges of the projects. Indeed, managing stakeholders has been identified as one of the most critical areas in project management (e.g. Crawford et al. 2006). Each project has multiple internal and external stakeholders on whom the project can have a significant impact or who can impact its performance (either positively or negatively). The form of impact the second group can bring to bear may well be a response to the way they are affected by the project. For this reason, it is important stakeholders are managed carefully throughout the project's life. This chapter discusses six classes of spontaneous stakeholders and nine classes of commissioned stakeholders in a project. In addition, we discuss a five-step stakeholder management process and propose two practical tools to support this process: a stakeholder register and a stakeholder report.

O. Zwikael and J. R. Smyrk, *Project Management*,
https://doi.org/10.1007/978-3-030-03174-9_5

From the literature

Box 5.1: Views of Stakeholders and their Management

Stakeholder management is widely accepted as an important area to enhance project success. While the definitions of "stakeholder" vary widely, they all capture the idea of someone "having an interest" in the venture. The Project Management Institute (PMI, 2017: p. 24) defines a stakeholder as "The people, groups, or organizations that could impact or be impacted by the project". El-Gogary et al. (2006) define stakeholders as "any group or individual who can affect or is affected by the achievement of the firm's/project's objectives". In some approaches, attempts are made to acknowledge different forms of stakeholding—such as those whose interest is "narrow" or risk-bearing (such as employees), investors, customers and community residents. Other definitions include "someone affected by a project and having a moral right to influence its outcome" (Bourne and Walker 2005). Stakeholders may include the funder, project team members, senior managers and employees from other departments of the performing and funding organisations.

Stakeholder management aims at assisting the key players in a project to identify, understand and influence the impact of external agents. Because stakeholders have such a significant potential impact (both direct and latent), several management frameworks have been developed and applied in various disciplines, such as e-business, managing environmental conflicts, software development and project management (Bourne and Walker 2005).

Different stakeholders may not only have dissimilar expectations of a project, they may even seek irreconcilable objectives. It is important, therefore, that the views of key stakeholders are identified and evaluated to assess potential threats and opportunities for a project (Kerzner 2017). The adjustments faced by some stakeholders because of the change is seen as a generic form of undesirable outcomes from a project, thus the "resistance to change" phenomenon (e.g. Ford et al. 2008; El-Gohary et al. 2006).

Zwikael and Smyrk (2015) emphasise the power of the project funder as a key stakeholder. In particular, because funders make the decision whether the project goes ahead or not, they can decide on the objectives for a project, even if some of the other stakeholders are not in full agreement with these.

5.1.1 *The Nature of a Stakeholding*

A *stakeholding* is the collection of all the interests in a project held by a *stakeholder*. A stakeholder can be an *individual* or a *group*. In this context, "an interest" can take either of two forms: the project has a potential impact on the entity, or the entity has a potential impact on the project. Such impacts can be either positive or negative. The management of stakeholders can have a significant impact on all aspects of the projects. Box 5.2 illustrates one such impact (on project scope).

Illustrative example

Box 5.2: Airport Expansion Project: How Stakeholder Management Influences Project Scope

Consider a project to increase the throughput of an international airport in which the initial scope was confined to airport-specific outputs such as a terminal, baggage handling system and new runway. Because later, the engineering analysis reveals that nearby residents would suffer significant increases in noise from aircraft movements, and it is decided that to engage these stakeholders, their homes should be noise proofed with double-glazed windows and sound insulation. Such a decision would increase the scope of the project (so that "modified houses" are now in scope). This provides an example of the complex interplay between different elements of a business case that must be considered during initiation.

Stakeholders are of a concern to key players because they can, in some cases exert a significant influence over the way a project unfolds. In general, a project will involve some entities who would be (at least in their own view) better off if the exercise did not proceed and who may, therefore, expose the venture to additional risk. In general, as risk increases, the attractiveness of a project to a funder decreases.

The interests of stakeholders ultimately stem from the changes wrought by a project. The accepted response to this problem is framed in terms of "managing change" with a focus on those whose stakeholding is related to their need to adapt to change. Rather than *managing change* per se, we seek to address this issue by *managing stakeholders*. The key players in a project seek to increase a project's attractiveness by managing stakeholders through programmes of stakeholder engagement. It should be noted that even those stakeholders who are opposed to the project are also considered for a programme of engagement. In cases where opposition takes the form of hostility, "engagement" may well be interpreted to mean "containment".

In the airport example above, the residents of the surrounding suburbs are stakeholders whose stakeholding can be summarised as:

- Potential impact of the project on the residents: reduced quality of life due to the increased noise of expanded aircraft operations.
- Potential impact of the residents on the project: generation of a groundswell of active opposition to the project.

While in this example, the local residents have a stakeholding defined by just two impacts, in general, the stakeholding of a particular entity could be defined by any number of impacts (positive or negative).

5.1.2 Spontaneous Versus Commissioned Stakeholders

In general, stakeholders emerge in a project for either of two reasons: an impact between the project and an entity arises spontaneously or an impact arises because the entity is commissioned to fill a role in the project.

Consider the following illustration. A project is undertaken to reduce local traffic congestion by extending a freeway, but this involves significant reduction of the local community's access to nearby botanical gardens. Amongst this community are active supporters of the gardens who: are well-organised, have established a large "fighting fund" and include many people with strong political connections. While (at the moment) they are not opposed to the freeway per se, they find it unacceptable that the gardens will be effectively isolated from most of the local roads. At the same time (although this is not yet known to the project proponents), a particular modification to the freeway design would alleviate the bulk of the community's concerns. Furthermore, the costs of such a change would still leave the project highly attractive to the funder. In such a situation, the project proponents would be well-advised to:

- Recognise the residents as stakeholders in the project.
- Understand the nature of their stakeholding.
- Set a strategy and establish a programme of actions for dealing with their concerns (which, in this case, might involve a change in project scope—to include a local access road).

In this example, supporters of the botanical gardens are spontaneous stakeholders regardless of any roles that might be recognised in the (eventual) governance model for the project. By way of contrast, the person who is eventually appointed as project manager becomes a commissioned stakeholder by virtue of that appointment—regardless of any spontaneous interest in the project.

Spontaneous impacts can arise at any point in the life of the project:

- Those householders whose properties adjoin the construction site and who will suffer from dust and noise are spontaneous stakeholders because of the (negative) impact on them arising from the work of constructing the freeway.
- Those motorists who use the completed freeway will experience reduced travel time, a (positive) impact associated with achievement of a (presumed) travel time target outcome. These motorists are also spontaneous stakeholders.
- The Botanical Gardens Trust (as an organisation) is a spontaneous stakeholder because of an undesirable outcome. If the freeway is completed to its original design, then the patronage will fall because of reduced access to the site.

When identifying stakeholders, the primary focus is on this "spontaneous" classification because their interest cannot be "switched" on/off at will. By way of contrast, we can virtually do that for commissioned stakeholders—by creating/terminating an appointment.

More formally, a spontaneous stakeholder is defined as an entity who has an interest in the project *regardless of any role that they might be commissioned to fill*. We propose a taxonomy of six generic classes of spontaneous stakeholders in a project:

1. **Funders**: those entities who have discretionary authority over the funding/deployment of the resources that will be released to the project. (The state government in the freeway project example).
2. **Beneficiaries**: those entities who are identified by the funder to receive a "flow of value" (i.e. a "benefit") from achievement of target outcomes from the project. (Motorists).
3. **(Positive) impactees**: those who experience a fortuitous "gain in value" at any time during the conduct of the project—regardless of the achievement of target outcomes. Take the case of a mobile canteen based in the general vicinity of the freeway construction site. The operator of the service may well become a positive impactee because of increased sales to the workers. (The operator is not, however, identified formally as a beneficiary because the funder is not seeking an increase in canteen sales as a return on the investment on the freeway). (Local businesses).
4. **(Negative) impactees**: those who experience a "loss of value" (i.e. a "disbenefit") at any time because of the project. (Neighbours).
5. **Customers**: those who will utilise the project's outputs and, in so doing, generate target outcomes. (Motorists again).
6. **Influencers**: those who, by virtue of their position or standing, are able to carry a significant body of opinion about the project. (The local newspaper).

These classifications are useful because one can postulate modes of behaviour for each class that allow predictions to be made about the response of certain entities due to different project scenarios. For example, negative impactees who have indicated implacable opposition to the project and who have become totally committed to its failure would be disqualified from joining a steering committee regardless of how well qualified they may be to fill that role. Such predictions can then help in the formulation of engagement programmes. Amongst other things, stakeholder analysis reveals to which of these classes each stakeholder belongs.

Note that in certain cases, a specific entity may be a member of more than one of these spontaneous stakeholder classes. Take for example, a resident who lives next to the proposed construction site and who will also have travel time reduced when the freeway is completed. That person is both a negative impactee and beneficiary.

More formally, commissioned stakeholders are defined as those entities whose interest in the project is a direct result of their appointment to play a formal role in the conduct of the project. We propose a taxonomy of nine generic classes of commissioned stakeholders in a project:

1. **Champion**: the person who drives preparation of the business case. The project champion is obviously a candidate for eventual appointment by the funder as *project owner*. (If a project owner is appointed from the outset, then the role of champion is unnecessary).
2. **Project owner**: a person who is held accountable by a funder for the realisation of target outcomes.
3. **Steering Committee members**: the steering committee is a small group of powerful project supporters who oversee the project's execution and eventually, outcomes realisation. Steering committees are discussed in more detail in Chap. 4 (project governance).
4. **The project manager**: the person held accountable by the project owner for delivery of the project's outputs.
5. **The project administrator, members of the project team and members of the project control group)**: those engaged (under the direction of the project manager) in the work of: planning the exercise, producing project outputs and managing that work.
6. **Reference group members and advisers**: reference groups are created to provide *collective* specialist input to the project, whereas advisers are *individuals* who are appointed to provide such services.
7. **Project assurance counsellors**: specialists who are appointed by the funder, owner or steering committee to ensure that the project is being run in accordance with accepted practice. (The role of *project* assurance is strictly above-the-line and hence is to be distinguished from that of *quality* assurance—which is strictly below the line).
8. **Probity counsellors**: engaged to ensure that the commercial dealings between the project and other parties are conducted in accordance with relevant guidelines, rules, regulations and the law.
9. **Suppliers**: those who provide goods/services to the project. Suppliers can be internal, external, commercial or non-commercial.

It is also worth noting that some misunderstandings about the nature of project stakeholding have given rise to flawed practices. Because, amongst other things the steering committee is involved in decisions about project scope, the appointment of a supplier to a steering committee could create a conflict of interest and so must be avoided.

There are just two situations in which an entity can be both a spontaneous and commissioned stakeholder—whereby they will appear in both the stakeholder register and in the project governance model.

- Someone who is commissioned to fill a role in the project governance model *just happens* to be a spontaneous stakeholder. In other words, such a spontaneous stakeholding *was not a precondition* for appointment of the entity to fill the commissioned role. It is critical to note that the person is being commissioned as a required project resource—not *because they have a spontaneous stakeholding*. In other words, their appointment is completely independent of their spontaneous stakeholding.

- A member of a reference group (or an adviser), on the other hand, is appointed to fill that role *because they are a spontaneous stakeholder*. In such cases, the appointment is made with the specific objective of engaging them in the project.

The first of these two situations may, under certain conditions, give rise to a conflict of interest whereby the role of the commissioned stakeholder is compromised by the nature of his/her spontaneous stakeholding. Such a conflict should be acknowledged in the issues register (introduced in Chap. 6) and resolved appropriately by the project owner (in consultation with the probity adviser if one has been appointed). Resolution may even require that the appointment be terminated.

5.1.3 Three Critical Characteristics of Spontaneous Stakeholders

Commissioned stakeholders are appointed for particular roles in the project and hence do not require a separate engagement programme. Spontaneous stakeholders, on the other hand, may need to be managed carefully to ensure that their interests do not expose the project to any unnecessary risk.

The rationale for including some type of stakeholder management element in frameworks of project management can be found by noting that stakeholders in a typical exercise display wide ranges of characteristics across three areas:

- **Their attitude** towards the project (expressed in terms of the views that they hold about the desirability of the initiative), ranging, in certain cases, from wholehearted support through to passionate opposition.
- **The power** that they can exercise over factors that may impact success.
- **Their influenceability** to change their attitude towards the project.

The first two of these areas (attitude and power) can be combined into one characteristic which we call *disposition*. Influenceability is a measure of how readily stakeholders may change their disposition.

Take two versions of the freeway project discussed above. "A" and "B" are identical in every respect, except for the patterns of support from one particular stakeholder group—local residents. "A" has very weak community opposition, while "B" has very strong community opposition. Project "B" is less attractive than project "A" because its higher level of opposition suggests a higher risk of failure.

In the second scenario, two versions of the same project ("C" and "D") are also identical in every respect. Furthermore, assume, for simplicity of discussion, that the numbers of supporters and opponents are equal, however, the levels of power that can be wielded over the project environment by the two communities of stakeholders differ between the two projects. In Project "D", the supporters are the most powerful, while for Project "C" the supporters have less influence. Again, it is clear that Project "C" is less attractive than project "D" because, although numerically the same, its opponents are relatively more powerful than its supporters.

Table 5.1 The stakeholder management process in outline

Subprocess	Output
Stakeholder identification	A list of stakeholders
Stakeholder analysis	Comprehensive details about each stakeholder
Engagement programme formulation	Stakeholder engagement programme
Engagement programme implementation	An implemented stakeholder engagement programme
Stakeholder engagement monitoring and control	Updated stakeholder engagement programme Updated stakeholder register

Disposition (and hence attitude and power) determines a stakeholder's potential impact on the project. Influenceability determines the ease with which a project's key players can get stakeholders to change their disposition. Knowledge of all three factors then enables a decision to be taken about appropriate levels of stakeholder engagement.

5.2 The Stakeholder Management Process

A regularised procedure to guide the work of managing spontaneous stakeholders is called a stakeholder management process (which is executed regularly over the course of the project). Commissioned stakeholders are not made the subject of this process because their stakeholding is "automatically" managed through their role description. The process is documented in a device called the stakeholder register—which is used to record the results of stakeholder analysis. At a high level, the stakeholder management process can be broken out into five sequential subprocesses, as described in Table 5.1.

1. **Stakeholder identification**. Stakeholder management begins with the creation (or updating) of a list of all those entities who have an interest in the project.
2. **Stakeholder analysis**. This activity involves an exploration of the two-way impacts between entity and project—leading to the assembly of a list of critical pieces of information about the entity and an assessment of the importance of the stakeholder as part of an engagement programme.
3. **Engagement programme formulation**. Here, based on the previous assessment, a decision is taken about how to engage those stakeholders who have been identified as having high importance in the previous step. Not all stakeholders warrant an engagement programme (which takes the form of a collection of actions to be undertaken during the project).
4. **Engagement programme implementation**. Engagement programmes are implemented by undertaking all the actions agreed in the previous step.

5. **Stakeholder engagement monitoring and control**. The progress of the engagement programme is monitored and adjusted as required.

All the information generated by the stakeholder management process is recorded in the stakeholder register (see Sect. 5.3.1 below).

5.2.1 *Stakeholder Identification*

The stakeholder management process begins with the identification of an entity who has a potential interest in the project. There is no regularised procedure for doing this. It could, for example take the form of a structured discussion involving all those who know something about the project. It could equally be triggered by the emergence of a stakeholder during other work on the project. The first step in the process is to confirm the name of the entity being considered as a stakeholder. It is important to note that the entries here identify *entities*—not the *form of stakeholding held by the entity*. For example, an entry "Beneficiaries" would not be appropriate because it does not identify the entity that has been identified as a beneficiary. Either of two forms of the label can be used to identify entities who are potential stakeholders:

- As an individual (by name): "Fred Nurke"
- As a group: "the residents of surrounding suburbs".

At any point in the project, the current list of entities who have a stakeholding in the project should be as comprehensive as possible, especially by including all those who are (or perceive themselves as being) negatively impacted by the exercise.

5.2.2 *Stakeholder Analysis*

When an entity has been identified as having a potential interest in the project, some simple analysis will quickly confirm the nature of that entity's stakeholding (if any).

While stakeholder analysis is often undertaken without the direct participation of the stakeholders concerned, in particular cases it may also be appropriate to involve them directly. Face-to-face discussion during analysis may encourage stakeholders to actively and openly discuss their attitudes towards the project, allowing more effective programmes of engagement to be formulated. Stakeholder analysis involves the following activities:

1. **Definition of the entity's stakeholding**. This activity has two parts: identifying (as a list) the nature of the potential impacts of the entity on the project (if any) and identifying (again as a list) the nature of the potential impacts of the project on the entity (if any).
2. **Classification of spontaneous stakeholding**. This involves listing all the forms of spontaneous stakeholding that apply to this entity (in light of the previous

step). These are drawn from the master list of nine generic classes of spontaneous stakeholders in a project (provided in Sect. 5.1.2).

3. **Identification of any consequences** (arising from this stakeholding) which go beyond those implied by the generic classes to which the entity belongs—for example, a project opponent has strong political connections. The only consequences that should be recognised here are those which have two characteristics:

   - they are not implied by the classes of stakeholding to which the entity belongs
   - the issue is relevant to the way project execution will unfold.

   Obvious consequences, such as a negative impactee being unhappy about the project and likely to foment opposition, are implied by the definition of a negative impactee, and so should not be identified here.
4. **Assessment of stakeholder importance**. Here, values are assigned for each of the three characteristics (attitude, power and influenceability) according to the judgement of whoever is filling the role of the stakeholder manager—in consultation with other key players. Using the simple algorithm outlined below, a value for *importance* (high, medium, low) is then derived from those three variables.

   - **Attitude**: a variable indicating the stakeholder's level of support/opposition for the project. Attitude has integer values in the range [−3, +3], where "−3" means strong opposition and "+3" means strong support. A higher absolute value of attitude is associated with a higher level of importance of a stakeholder.
   - **Power**: A simple three-value variable indicating the stakeholder's ability to influence the course of the project. Power has values of 1, 2 or 3 (corresponding to low, medium, high). A higher value of power is associated with a higher level of importance of a stakeholder.
   - **Influenceability**: A simple three-value variable indicating the ease with which the project's key players can change the stakeholder's attitude towards the project. Influenceability has values of 1, 2 or 3 (corresponding to low, medium, high). A higher value of influenceability is associated with a higher level of importance of a stakeholder.

   The product of all three parameters generates a score between −27 and 27. A large positive score indicates an important supporter, while a large negative score indicates an important project opponent. The focus of supporter engagement is maintaining that support, while for opponents it is fostering support. For practical purposes, values of stakeholder importance above 10 or below −10 are considered high; between −4 and 4 are considered low and anything else is considered as a medium.
5. **Setting objectives for engagement**. Here, an answer is provided to the question "What sort of situation are we trying to bring about by engaging this entity?" The objective set for engagement is determined by the "importance" index calculated above. Based on this, together with opportunities to influence each stakeholder, the project proponents can determine which forms of engagement are desired and feasible.

This analysis generates the information required to formulate an engagement programme, with an emphasis on stakeholders who receive high importance score—as discussed in the next section.

5.2.3 *Stakeholder Engagement Programme Formulation*

In light of what has been discovered during the analysis of each entity's stakeholding, an appropriate programme of actions ("a stakeholder engagement programme") can be assembled to engage those whose interests need to be managed. There are two outputs from engagement programme formulation:

- A stakeholder engagement programme which takes the form of the various entries scattered throughout the stakeholder register.
- A communications strategy which harmonises and reconciles the communications elements of these entries.

A useful approach to generate effective engagement ideas is to use the following three strategies (although ideas from other sources and strategies can also be useful for consideration):

1. **Communicate with the stakeholder**. An example would be to invite the stakeholder to a regular monthly briefing by the project manager. A corresponding entry in the stakeholder register appears under the heading: "Include in the communications programme". Such entries simply *identify* the proposed forms of communication such as "invite to monthly project briefings", "grant website access" or "provide with copy of regular newsletter". If it is believed that a stakeholder will be better off as a result of the initiative, but is unaware of that, then some form of "marketing" communication may be appropriate.
2. **Increase the project's value to the stakeholder**. To increase the value of the project to a stakeholder, it will usually be necessary to either add additional outputs (that go beyond those implied by the ITO model) or change their fitness-for-purpose features. The construction of new towns for people displaced by the Three Gorges Dam in China is a case in point. A corresponding entry in the stakeholder register would identify the proposed action as "Provide new permanent residential area for people displaced by dam impoundment". An addition to the scope of the project arising from stakeholder engagement is a form of a *non-ITO output*. Non-ITO outputs (which is discussed in Chap. 9) may arise for other reasons—such as risk mitigation or issues management. It is important to note that this strategy is based on the idea that the "true" underlying value of the project to "wavering" stakeholders must be increased (by changing the list of outputs and/or the lists of fitness-for-purpose features of the existing outputs). Naïve attempts to "sell" the project without making such changes may well result in an unhealthy level of cynicism about the exercise.
3. **Have the stakeholder participate in the conduct of the project**. A project to move office may have important implications for staff travel. In that case, it may

be helpful to form a reference group of staff to offer comments and provide input to the project team. Such entries are confined to the identification of the proposed action—such as "include in staff liaison reference group" or "invite as member of technical standards working group". The only ways in which a spontaneous stakeholder might be engaged through participation in the project is via an invitation to fill a role in the *reference* division. It would never be appropriate to appoint spontaneous stakeholders to fill any other project governance role *as a strategy for engaging them*—because candidates for those positions are selected according to their competence and proficiency. All supporting detail about the proposed role is addressed in the project governance model with a formal role definition. We can now revisit the earlier example of the project manager who, because of living next to the construction site, also becomes a spontaneous stakeholder (as well as a commissioned stakeholder). The person serving as a project manager has, however, been appointed because of his/her project management skills—not because of the spontaneous stakeholding.

The engagement programme for a particular stakeholder will *always* include some form of communication (the first of the three generic engagement strategies discussed above). If it is felt that an entity does not merit inclusion in the (eventual) project communications programme, then it would be safe to assume that the entity can be ignored. The other two generic engagement strategies may or may not be appropriate in particular instances.

In the freeway illustration discussed above, based on the three proposed generic strategies, the engagement programme for local residents may appear as:

1. Communicate with the stakeholder:
   - Provide background material on the project website.
   - Meet informally over coffee every month.
   - Include in the distribution list for the project newsletter.
2. Increase the project's value to the stakeholder:
   - Change the fitness-for-purpose features of the highway by including an on/off-ramp to the botanical gardens.
3. Have the stakeholder participate in the conduct of the project:
   - Invite to join a reference group made up of representatives of botanical garden supporters.

It should be noted that the entries in this example go into no detail about the website, the monthly meetings, the newsletter, the off-ramp or the charter of the reference group. The design, development and creation of such outputs are usually all undertaken during project execution, following planning (the project's second global phase).

Face-to-face contact may figure prominently in the programme assembled for particular stakeholders, but care has to be exercised when considering such an approach, especially with entities who are opposed to the project. Of particular note are situations related to "community action" and "special interest" groups where public meetings may play a critical role. Activities of this kind is challenging, but if done well, can go a long way towards ameliorating the worst effects of public antipathy. The consequences of poorly conducted public meetings can be catastrophic and so they must be planned, promoted, structured and professionally chaired.

5.2.4 Deriving a Stakeholder Communications Strategy from the Stakeholder Register

When addressing that part of the engagement programme relating to communications, it is necessary to integrate all the separate elements across the entire community of stakeholders into a cohesive *communications strategy*. Fragmented, conflicting and uncoordinated attempts at communication could have the opposite effects to those set as objectives in the original engagement programme. The communications strategy itself has two parts: a simple overview explaining how the various elements of stakeholder communication all fit together and a summary of each communications medium that will be employed (but again without any elaborating detail). The relationship between the stakeholder register and the communications strategy is this:

- The *stakeholder register* lists all the forms of communication that are to be adopted for each stakeholder. This means that references to, say, the project website will appear in a number of places. (Loosely, the stakeholder register summarises communications as two-level hierarchy—with stakeholders at level 1 and form of communication at level 2).
- The *communications strategy* inverts this hierarchy so that forms of communication are shown at level 1 and stakeholders at level 2. The rationale for this is that the communications strategy shows in one place all the stakeholders who must be given access to, for example the project website. This then helps everyone understand what the website is to do—as well as making the eventual planning of its development simpler.

The communications strategy is covered more thoroughly in Chap. 9 with a discussion of both its structure and the procedure for its assembly.

5.2.5 Engagement Programme Implementation

With the engagement programme in place (and summarised in the stakeholder register), the detailed planning of stakeholder engagement can now proceed, much of it as part of the "standard" work undertaken during the global project phase called planning. Details about all the work involved in stakeholder engagement, and in producing the outputs associated with stakeholder engagement (such as websites, newsletters and so on) are incorporated into the normal planning activity and eventually reflected in the work breakdown structure, Gantt chart and schedule of milestones (both introduced in Chap. 9).

In the interests of expediency, parts of the stakeholder engagement programme may well be implemented during the planning global phase, with the bulk of it undertaken (as expected) during execution—and much of that very early in the phase. The foundation for this work is the engagement programme, based on the three generic engagement strategies discussed above.

5.2.6 *Stakeholder Engagement Monitoring and Control*

To ensure that the stakeholder engagement programme is implemented successfully, whoever is filling the role of stakeholder manager (the project administrator, an appointed project team member or the project manager directly) has to monitor its effectiveness throughout the project.

The case study in Box 5.3 demonstrates how the suggested process has been implemented in a project to engage opposing stakeholders back into the project.

Illustrative example

Box 5.3: How Stakeholders can be Effectively Engaged
This case study concerns a road construction project, managed by the Greater Wellington Regional Council (GWRC) in New Zealand (Zwikael et al. 2012). It represents a proactive approach to the model described above, where key stakeholders take an active part of the analysis process.

The GWRC had been seeking to address the increasing problems of congestion, safety and community severance along the existing State highway route north of Wellington. The Western Corridor carries from 20,000 to 75,000 vehicles and 11,500 rail passengers per day. According to the GWRC, there was an urgent need of an affordable, safe, efficient, reliable and sustainable Western Corridor transportation network that provides reasonable capacity. In this context, the GWRC proposed a Western Corridor Project, which included public transport improvements, travel demand management initiatives and a staged programme of road improvements that address safety, reliability and capacity.

Some stakeholders supported the project while others voiced their opposition towards it. The conflicting views among stakeholders presented increasing challenges to the transport managers of GWRC. In response, a stakeholder engagement programme was assembled. These stakeholders were successfully engaged in a process based on the following steps.

Those with an interest in the Western Corridor project were identified by GWRC managers. They include internal stakeholders (such as the Finance Department, and GWRC transportation managers), as well as external stakeholders (such as the government and "green" representatives).

Representatives from all stakeholder groups were invited to participate in a "raising issues" session. The stakeholders who attended the session generated a total of 72 issues, each related to an opportunity or obstacle, based on their opinions on the project, such as "cost of accidents", "alternative use of money" and "environmental damage". The stakeholders then grouped issues that were related to form clusters. A descriptive name has been given to each cluster. In this exercise, the stakeholders developed 12 such clusters. For example, the "cost" cluster involved eleven out of the 72 issues, related to aspects such as the high cost of the project, sources of funds, alternative use of the money and cost of accidents. Stakeholders then identified variables associated with each of these clusters. For example, amongst the 21 variables included in the "cost" cluster, were "Cost of congestion", "accidents", "number of days road is closed due to hazards" and "travel time". All variables were then linked to generate a "causal loop model". For this purpose, stakeholders tried to establish relationships between all variables. They identified pairs of related variables—connected by an arrow.

In this process, stakeholders were able to analyse the relationships among the following variables: "average number of trips per person per day", "accidents", "number of days the road is closed due to hazards", "social impact on community", "support of community stakeholders", "support of political stakeholders", "Western Corridor project", and "change in trip volume and distribution". At the end of this exercise, all agreed that this model represented a shared view of the exercise. Understanding the interest of different stakeholders in this project, and its potential impact on each of them also contributed significantly to the development of the project business case.

5.3 Stakeholder Management Tools

5.3.1 The Stakeholder Register

The primary tool in stakeholder management is the stakeholder register, which is used to document the results of analysis and outline the form of engagement being proposed for each project stakeholder. The stakeholder register appears in the business case and later in the project plan. Because it is a register, this tool takes the form of a table where rows are associated with entities) and columns with attributes. (We recognise three registers in our approach to project management—the other two related to risks and issues). A set of attributes making up a typical stakeholder register would include columns for:

1. **Stakeholder ID**: typically, "S01", "S02" and so on.
2. **Entity name**: the name of the *entity* being considered as a stakeholder.
3. **Description of entity**: useful when this is not obvious from the entity name.
4. **Status**: An entry in the stakeholder register develops over time—as the stakeholder management process unfolds. When first identified, only a name and description would normally be known, while by the end of the project it will be closed. This evolution is usefully summarised with a status flag drawn from the following sequence
    - Identified: only a name and description of the entity are available.
    - Prepared: an engagement programme for the stakeholder has been proposed but not yet implemented.
    - Active: an engagement programme for the stakeholder has been implemented and is now being monitored.
    - Closed: no further action is required to engage the stakeholder.
5. **Stakeholding**: lists of the potential impacts of the project on the entity and the potential impacts of the entity on the project.
6. **Classes of "spontaneous" stakeholding**: a list of the classes to which this stakeholder belongs:
    - Funders
    - Beneficiaries
    - (Positive) impactees
    - (Negative) impactees
    - Customers.
    - Influencers
7. **Consequences (associated with this stakeholding)**: This entry is confined to noteworthy consequences that are not obvious from the forms of spontaneous stakeholding.
8. **Attitude**: simple three-value variable indicating the stakeholder's level of support for/opposition to the project.
9. **Power**: A simple three-value variable indicating the stakeholder's ability to influence the course of the project.
10. **Influenceability**: A simple three-value variable indicating the ease with which the project's key players can change the stakeholder's attitude towards the project.
11. **Level of importance (of stakeholder)**: Calculated as the product of the values provided for the last three variables. (Guidance on the interpretation of this score is provided above in Sect. 5.2.2 step 4).
12. **Objective of engagement**: a clear, simple statement of the sort of situation sought from the engagement of this entity.

13. **Engagement programme**: Identifies the mechanism(s) that will be employed to engage this entity appropriately in the project. In general, this will include some form of communication during the project.
14. **Assigned to**: Person responsible for the engagement programme for this stakeholder.

There is considerable flexibility in the design of the stakeholder register by incorporating other columns such as anticipated duration of stakeholding. Table 5.2 provides an example of an entry in the stakeholder register for the Botanical Gardens project, including comments related to this particular exercise.

Table 5.2 An example of an entry in the stakeholder register

Stakeholder attribute	Stakeholder entry	Comment
Stakeholder ID	S42	
Entity name	Friends of the Botanical Gardens Inc.	This is a *group* rather than an *individual*
Description	A formally recognised body of volunteers who maintain and operate the gardens	This group has strong political connections and strong support in the local community
Status	Active	Engagement programme was approved by the project owner in the February steering committee meeting
Stakeholding	1. Potential impact of the project on the entity: Botanical Gardens will face reduced numbers of visitors	
	2. Potential impact of the entity on the project: They may conduct a programme of active opposition	
Classes of "spontaneous" stakeholding	Influencers and negative impactees	
Consequences (associated with this stakeholding)	They have strong political connections and could use these to force major change in the design.	
Attitude	−2	Attitude ranges between −3 and +3.
Power	2	Power has values of 1, 2 or 3.
Influenceability	3	Influenceability has values of 1, 2 or 3.
Level of importance	High	$-12 = (-2 * 2 * 3)$.
Objective of engagement	To reduce their level of opposition	
Engagement programme	Provide with access to the project website	Related strategy: Communicate with the stakeholder
	Meet informally over coffee every month	Related strategy: Communicate with the stakeholder
	Reconfigure on/off ramps to enhance access to gardens	Related strategy: Increase the project's appeal to the stakeholder

(continued)

Table 5.2 (continued)

Stakeholder attribute	Stakeholder entry	Comment
	Invite to join a "Botanical Gardens reference group"	Related strategy: Have the stakeholder participate in the conduct of the project
Assigned to	Christine Bossuat	Christine Bossuat is a very active member of Friends of the Botanical Gardens. She has particularly good interpersonal skills

Because the community of stakeholders tends to evolve slowly over time, changes to the stakeholder register are relatively infrequent, and so its maintenance is not particularly demanding. The project manager holds the primary stakeholder register, but there may be occasions when others hold "supplementary" registers. For example, the project assurance counsellor (when the role exists) may assemble a confidential stakeholder register when premature release of information about a project opponent would cause irreparable damage.

5.3.2 *The Stakeholder Report*

Because the stakeholder register tends to be relatively stable, not all entities demand regular detailed attention and monitoring. For this reason, it is not necessary to display the entire register in, for example, regular project status reports. Instead, only selected entries are highlighted for periodic review. Such an extract from the stakeholder *register* is called a stakeholder *report*. The stakeholder report has exactly the same structure as the stakeholder register—but usually with far fewer entries.

A stakeholder report is confined to those entities for whom changes have been made to the register since the last report. This means that a report shows only those entities who are new or lapsed stakeholders—as well as those whose entry in the register has been amended since the last report. Whenever a stakeholder report is tabled, it should be accompanied by a link to the full register—thus allowing readers who need more details to find it easily.

Summary
In this chapter, we have discussed the concept of a stakeholder—an entity who can impact the project and/or be impacted by the project. We distinguish spontaneous from commissioned stakeholders because each group is handled differently. Spontaneous stakeholders are the subject of an engagement programme (and so appear in the stakeholder register), while commissioned stakeholders are appointed (and so appear in the project governance model). Spontaneous stakeholders are ranked by importance (based on judgements about each of three underlying characteristics) and made the subject of a formal management process. All the relevant information about a project's stakeholders and their management is summarised in a stakeholder register.

Chapter 6
Risk and Issues Management

Abstract In this chapter, we introduce the concepts of risk and issues, and then discuss tools to support their management during the project. The management of risk and issues are based on a common underlying process involving a sequence of steps including identification, analysis, programme formulation, implementation and control. The most important tools take the form of registers and reports for each of risk and issues.

6.1 Risk Versus Issues

A project is subjected to all sorts of influences over its life—many of which hinder progress towards achievement of its baseline documents. By adopting a formal structured approach, we can enhance our ability to identify, analyse and respond to these influences so that their negative impact on the project is reduced. The framework for doing this is based on a simple two-way categorisation of these influences as "risks" or "issues".

At first blush, risks and issues look very similar, but they each nevertheless require their own distinct management process. In particular, if a risk is realised, it evolves into an issue. Although a *realised risk* thus qualifies as an issue (because it has happened)—its management is approached differently. Once an entry appears in the risk register, it can never be "demoted" to the issues register (see Sect. 6.5.4.1). The reason is that the risk register already contains important details about the actions that will be taken if it is realised (In other words, there is no point to now moving all that detail out of the risk register into the issues). Issues are further distinguished from risks in that:

- An issue can take any of three forms—one of which is an event that has happened (As discussed below, issues can also take the form of questions or tasks).
- Unlike a risk, an issue is not necessarily associated with damage to the project.

All these observations about the difference between issues and risk can be summarised using a decision tree—as shown in Fig. 6.1. Risks are discussed in Sects. 6.2–6.4, whereas issues in Sect. 6.5.

O. Zwikael and J. R. Smyrk, *Project Management*,
https://doi.org/10.1007/978-3-030-03174-9_6

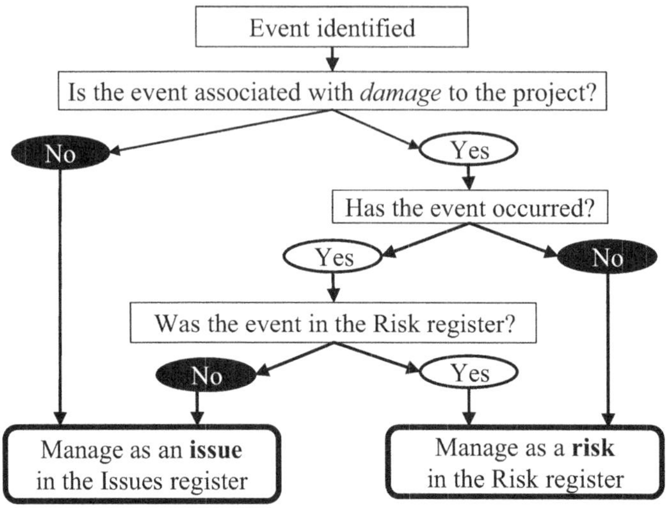

Fig. 6.1 Selecting the appropriate register to manage an event

6.2 The Nature of Risk

6.2.1 Risk and Uncertainty

Throughout this chapter, we use two closely related (but subtly different) terms—"risk" and "threat"—which bear further discussion. Threat is defined as a *possible future event* that would "damage" the project if it materialised. Although we formally define a risk with an entry in a risk register (discussed in Sect. 6.4.1 below), it already has a number of widely accepted meanings in common parlance—such as:

- A *probability* ("there is a 50% risk of rain").
- A *measure of uncertainty* about a result. This use accords with its interpretation in investment theory ("The risk-return diagram").
- *As a synonym for a threat* ("There is a risk that the concrete will fail a compression test").

As a matter of expediency, we use "risk" loosely—relying on the context to convey its meaning. Risk is associated with uncertainty about the eventual achievement of values for particular parameters that are set, estimated, derived or expected. These include target outcomes, undesirable outcomes, timeframe and cost. Uncertainty is only of concern to a project if it is associated with a *reduction* in the target worth of the project as declared in a business case. Worth is a concept representing the "value" of the project. Uncertainty arises from events that have two characteristics: they are stochastic (that is, they may or may not occur) and if they do occur, the project will be "damaged" (by having its target worth reduced). Such events are called threats. The potential resignation of a key team member is an example of a threat to a project. By implication, if an event can cause no reduction in a project's worth, then it does not represent a threat to that project. In other words, risk can be viewed as *uncertainty about the achievement of a project's target wor*th.

Risk is a phenomenon of the future, not the past, and so while a prospective project will have measurable level of risk, for a completed project this will be zero or (in many cases) "close to" zero. All major risks will have evaporated by the time the end of the project is reached—leaving only those associated with uncertainty about the way that the ongoing operational environment may unfold in the future.

Since the multitude of threats that a project faces during its life vary widely in their significance, we need a method to make systematic judgements about the importance of each. This will enable threats to be ranked so that attention can be directed at the most important, while the least important may not merit any attention at all. Threats can be ranked by their *expected damage*, a qualitative measure of their "importance" as discussed below.

The overall "riskiness" of a project is gauged by the spread of the possible damage arising from all the threats to which it is exposed. We provide a more formal statistical treatment of riskiness in Chap. 7.

Consider two projects A and B that are identical in all respects except that the riskiness (the spread of possible levels of damage from various threats) of B is greater than that of A. As the riskiness of a project increases the funder can expect a correspondingly increase in the range of possible values of worth that might be experienced, however, our assumption of satisficing behaviour (see Sect. 6.2.2 below) implies that avoidance of underperformance is more important than achievement of over-performance. In other words, despite the fact that increased riskiness brings with it the possibility of higher returns, the funder will be conscious of the fact that the possibility of poorer returns also increases. This implies that project A is a more attractive investment opportunity than B. If projects are to be ranked in terms of their attractiveness for funding, then, in addition to the worth declared in the business case, we must also consider their riskiness.

While this overall approach is consistent with the accepted principles of investment theory, most of the analytical tools used in that discipline cannot be applied directly to project funding decisions for two reasons:

- We are, in general, limited to qualitative (rather than quantitative) measures of worth and riskiness because empirical data is, in general, not available. Qualitative measurement is based on values of variables that are *judged*. By way of contrast, quantitative measurement is based on the values of variables that are *measured*.
- We are concerned only with *downside* risk—as discussed in the following section.

6.2.2 Downside Versus Upside Risk

In investment theory, risk is seen as the uncertainty of *exceeding*, as well as *falling short of* some expected value. Here, we are concerned only with falling short of a threshold, a stance that is consistent with satisficing (rather than optimising) behaviour (see Box 7.3). We identify this particular approach to risk, (in which we are concerned only with the possibility of *not* reaching the thresholds that define the project's worth), as *asymmetric* risk management. This position is based on an assumption—that the behaviours of a project's key players are driven by *satisficing* rather than *optimising* goals. Amongst other things, this enables us to treat risk as intrinsically undesirable—which has important implications for the tools used in Chap. 7 (where we discuss project attractiveness).

Because of the subtle difference in the way that risk is handled here and in financial investment, we use the word "riskiness" instead of "risk".

6.2.3 The Event-Impact Model of Risk

All forms of risk can be expressed as specific instances of an event-impact model, as illustrated in Fig. 6.2.

The event-impact model of a risk is a mechanism involving four time-related components:

1. A threat (in the form of a triggering event), such as, "Project manager resigns" (see Zwikael and Ahn 2011). While the events to which threats are related can be either above- or below-the line, typically, most are below-the-line (associated with the production of a project's outputs).
2. A chain of subsequent events, each caused by the one before. These are peculiar to each threat and require no explicit description—they are shown here simply for completeness.
3. A consequence (in the form of an undesirable situation). For example, "if the project manager resigns there will be no one capable of concluding negotiations with the subcontractor". It is only necessary to articulate the consequence of a threat if the cause–effect relationship between the triggering event and its eventual impact is not clear.
4. A (damaging) impact, which lowers the worth of the project.

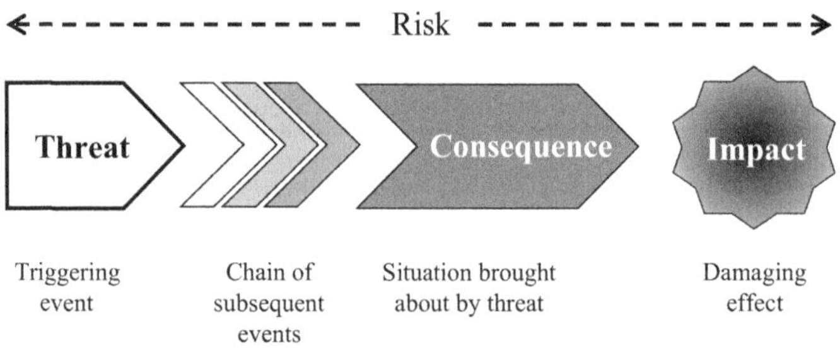

Fig. 6.2 The event-impact model of a project risk

The appearance of the word "Risk" above the arrow in Fig. 6.2 signals that our formal definition of risk relates to an entire occurrence of the even-impact model (not just to the emergence of the threat).

Threats have a *likelihood* of occurrence which can be viewed as the level of our belief or expectation that the event will happen. Likelihood is a form of Bayesian probability which must be judged qualitatively based on our knowledge of the threat—unlike frequentist probability, which is estimated empirically from data. Accordingly, likelihood also has a value in the interval [0, 1] where the selected value indicates the level of our belief or expectation that an event will occur. If we assign a value of "0" to the likelihood of an event "E", we are effectively declaring our belief that "E" cannot happen—and so we would then be justified in ignoring it in our analysis of a project's risks.

Likelihood (Li) can be viewed as the inverse of the "level of surprise" that someone would experience on discovering that a threat had been realised. This relationship can be exploited to help set an appropriate value for likelihood—as discussed in Box 6.1. Likelihood will be moderated by the information about the threat that is available. For example, a project owner who became aware that the project manager was very unhappy, may well form a view that his resignation was "highly likely" and set its value at, say, 0.75.

Practicalities

Box 6.1: Using "Level of Surprise" to gauge Likelihood

The level of surprise we would experience if a threat is realised indicates the likelihood of that threat occurring. By conducting a simple thought experiment, this relationship can be used to help select the most appropriate value for the likelihood of a particular threat.

Consider the threat "Prime contractor declares bankruptcy (within the next three months)". Imagine that, as you walk into the office one morning over the next three months, a team member announces that "Our prime contractor has gone bust!". Your level of surprise on learning this will be inversely related to the likelihood of the threat. If your reaction would be one of incredulity (a high level of surprise) then your judgement of the likelihood of such an event would be closer to 0 than to 1.

By way of contrast, consider a scenario in which you have read in the business press that the contractor is under significant financial pressure. The thought experiment has you concluding that, in light of what you know, such a piece of news would not surprise you in the slightest, and so you set a value for likelihood closer to 1 than to 0.

A damaging impact is gauged by its "severity" (Se)—a qualitative measure of the total reduction in the target worth of the project if the threat were to be realised. While in many cases, it would be possible to use mathematical and financial modelling to estimate the reduction in worth caused by a threat, this is not always possible in practice because of constraints on time and resources. Severity is usefully gauged in terms of the same variable used to gauge worth. In those special cases where worth is measured in dollars for example, then Se would also be gauged in dollars.

Given the likelihood of a threat and the severity of its damage to the project, we can calculate the *expected damage* of that threat as Li*Se (the product of likelihood and severity). Expected damage serves as our gauge of the level of importance attached to a threat.

Practicalities

Box 6.2: Gauging Severity

Severity is gauged as the loss of a project's worth if the associated threat is realised. Worth ultimately derives from target outcomes, undesirable outcomes and costs. A threat can impact any these factors unfavourably—each impact resulting in an incremental loss of project worth. Because, in general, the units of measurement for target outcomes, undesirable outcomes and costs will be incommensurate, estimates of the losses attached to a threat take the form of qualitative judgements.

To make a meaningful estimate of the loss of project worth arising from the realisation of a particular threat, the following systematic procedure is suggested:

1. List all: target outcomes, undesirable outcomes and costs.
2. For each threat:
    - Take the items from the list created in the previous step, one at a time and judge the magnitude of the associated loss.
    - Judge the overall loss of worth based on this list of losses—which then becomes the severity of the damage arising from realisation of the threat.

Risks are time-dependent in two ways. First, the timing of the triggering event may determine the severity of the impact. For example, the impact of a prime contractor declaring bankruptcy may be much more severe if it happens early in execution than, say, when execution is almost finished. Second, the likelihood of a triggering event will depend on the horizon being considered. For example, the threat "Key technical adviser leaves before the end of the week" will, in general, have a lower likelihood than the threat "Key technical adviser leaves before the end of the month". The first of these two phenomena implies that entries in the risk register must be reviewed regularly—to ensure that they are up to date. The second indicates that the names and descriptions of threats must be assembled thoughtfully.

Two important forms of risk are distinguished by the nature of their triggering events, which are either discrete or continuous. The threat "Prime contractor declares bankruptcy" is a discrete-variable event (because it either happens or it does not). By contrast, the threat "rain falls" is an example of a continuous-variable event because rainfall can range from being merely irritating (a single light shower) through to catastrophic (causing severe flooding).

The event-impact model handles discrete risk quite well, but continuous risk rather poorly. Despite this, while not particularly elegant, the model can be adapted to handle continuous threats. In the case of "rain falls", a number of representative values from the rain continuum would be handled as discrete events. For example, one could consider three representative events: a light shower, moderate rain, a deluge—and handle each one separately. In doing so, the limitations of this approach must be borne in mind when using it to gauge the "riskiness" of a project.

6.3 The Risk Management Process

The significance of project risk management arises from the following observations:

- Risk reduces the *attractiveness* of a project which is discussed more formally in Chap. 7.
- Most projects are exposed to chance events that have damaging consequences of one form or another. The damaging effects of these events can, in extreme cases, lead to project failure.
- Many events of this kind are identifiable in advance.
- The damaging events to which a project is exposed vary in terms of their expected damage and can be ranked accordingly.
- Actions against certain threats are available which have the effect of reducing their expected damage.
- The costs of taking action against some risks appear low when compared with the extent to which their expected damage is reduced.

Under these conditions, we appear *compelled* to act against the risk (One could even argue that a failure to act would constitute a form of professional negligence). The approach described below has two characteristics: it is peculiar to the *project* (as distinct from the *operational*) environment and it is concerned only with downside risks (as discussed in the next section).

As discussed in the next two subsections below, the overall risk management process is made up of:

- A threat management process for each particular threat.
- A risk control process to ensure/confirm that all threats are satisfactorily managed.

6.3.1 Managing Threats

The threat management process is triggered by the emergence of a new threat and involves a sequence of four subprocesses—as summarised in Table 6.1.

Note that, we use the word "mitigation" for *any* action that will reduce the importance of a threat—regardless of whether this action impacts likelihood or severity.

This process is led by whoever is recognised as the risk manager (a role introduced in Sect. 4.1.10). The threat management process is ongoing throughout the entire project, starting very early during the initiation phase, (when some threats to the project are first identified) and continuing through to the end of the exercise. The results of each activity are documented in a device called the risk register—discussed below.

The approach to threat management is based on a simple principle of decision analysis: a course of action should only be adopted if its anticipated effectiveness more than compensates for the estimated costs involved in its execution. In accordance with that principle, we propose the following broad methodology:

1. Identify a threat to which the project is exposed.
2. Gauge its importance in terms of its expected damage (in the absence of any action to mitigate the threat).
3. Decide if the level of expected damage merits the consideration of some form of mitigation. If the expected damage is too small, no further action is required.
4. Assemble a list of specific potential mitigating actions to deal with the threat.
5. Estimate the costs of each action in the list.
6. Gauge the change in the importance of the threat attributable to the list of proposed mitigating action. This change (in expected damage) is a measure of the *effectiveness* of the list of mitigating action.
7. Using the attractiveness map (see 7.3) for the project, decide on accepting/rejecting the proposed list of mitigating actions. Section 7.3.3 discusses this decision in more detail. The methodology has a number of immediate implications:

    - Expected damage must be assessed twice: first on the assumption that risk mitigation actions are not implemented, and again assuming that they are.

Table 6.1 The threat management process in outline

Subprocess	Outputs
Threat identification	A list of threats
Threat analysis	Details about each threat
Threat mitigation programme formulation	A threat management programme
Threat mitigation programme implementation	An implemented threat management programme
Threat monitoring and control	An updated threat management programme

To distinguish the two measurements, we refer to likelihood, severity and expected damage in terms of their values relating to the situation where no risk mitigation is in place ("pre-"), versus their values relating to the situation where a risk mitigation programme is in place ("post-").

- The cost of the proposed programme of mitigation must be estimated.
- Some threats do not merit risk mitigation. The others may retain a certain level of post-expected damage after all mitigating actions have been implemented. Together, both classes impose what is called a *residual risk* on the project.

6.3.1.1 Threat Identification

Threat identification is simply the process of identifying and describing events that could harm the project. A threat can emerge during any of a project's four global phases—from initiation through to outcomes realisation. This is best done by small groups of people who understand relevant aspects of the project environment. The risk registers of past projects can be a valuable source of potential threats for new projects. We propose a particular word structure for threats, taking the form of "newspaper headline style statements", that is, they are expressed as they would appear on the billboards of a newspaper the day after they occurred.

"Project manager leaves", "groundwater found to be polluted", "state government fails to pass enabling legislation", "new process proves slower than anticipated" and "prime contractor declares bankruptcy" are all examples of this format.

It is important to note that *any* threat causing the worth of a project to fall is recognised—including those related to any regular future business operations—in particular, those conducted during the global phase of outcomes realisation. As discussed in Chap. 2, utilisation is always in the form of the execution of some downstream processes. A threat to a downstream process will result in some sort of impediment to its execution—and hence degrade one or more performance metrics. Our principle of duality says that a target outcome always takes the form of a metric attached to a downstream process. Conclusion? Any threat to a downstream process must be recognised as a project threat.

6.3.1.2 Threat Analysis

Threat analysis has three parts—each concerned with setting a value for one of the parameters of a threat:

1. **Likelihood** is usefully expressed as a probability—with a value in the interval [0, 1].
2. **Severity**, (the damaging impact of the threat if it is realised) is gauged as the amount of the project's target worth that would be lost if the threat arises.
3. **Expected damage** reflects the importance of each threat and is gauged as the product of likelihood and severity. By ranking all risks in terms of their expected

Table 6.2 The six forms of damaging impact that project worth can suffer as the result of a threat

Worth variable	Magnitude	Timing
Benefits	Reduced	Delayed
Disbenefits	Increased	Advanced
Costs	Increased	Advanced

damage, the most important can be isolated for a mitigation programme of some kind.

The damaging impact of a threat takes the form of a fall in the project's worth. Since worth is a function of three variables (benefits, disbenefits and costs—all introduced in Sect. 2.5.5 and discussed in Chap. 7), expected damage must be reflected in an adverse movement in one or more of these three variables. Furthermore, there are only two ways in which each worth variable can move unfavourably:

- In magnitude, for example costs *increased*.
- In timing, for example benefits *delayed*.

The combination of two forms of impact (magnitude and timing) and three project worth variables (benefit, disbenefits and costs) generates six generic forms of damage that a project can suffer—as summarised in Table 6.2.

The values assigned for the severity of a project's damage are drawn from the scale used to gauge worth. Selection of an appropriate value for severity involves two steps: the first is to identify which combination of the six forms of damage applies to the threat under analysis, the second is to make a judgment about the magnitude of the associated damaging effects on worth. For the threat "the project manager leaves" under normal conditions, only two forms of damage would be expected from amongst the entries in Table 6.2: benefits delayed and costs increased. At this point in the process, "values" have been decided for both likelihood and severity, thus allowing an expected damage to be calculated (as their product).

The calculation of expected damage (as the product of likelihood and severity) adopted here uses a *grading rule* (by which the importance of a threat can be "graded"). There are other approaches to grading a threat based on its likelihood and severity. One of these uses a *grading table* in which values for likelihood and severity appear as row and column headings, with the cells indicating the corresponding value for expected damage. Such a device is suited to situations where importance, likelihood and severity are all measures using *ordinal* variables. The "Probability and impact matrix" appearing in the PMBoK is an example of a grading table.

6.3.1.3 Mitigation Programme Formulation

Risk mitigation is based on actions that lower the expected damages of selected threats. Such actions are of two kinds:

- Preemptives: which reduce the likelihood of the threat emerging in the first place.

- Contingencies: which reduce the severity of the damaging impact if the threat is realised.

A risk mitigation programme is a made up of all the agreed actions that seek to reduce the expected damage from all threats in the risk register, but it contains no details about those actions. For example, it might be decided to formally appoint and train a deputy, as one of the contingencies for the threat "Project manager leaves". At this point, there is no detail about the proposed training programme and so the next step in risk management sees the assembly of a detailed plan for implementation of the programme. The project manager normally administers the process, but execution of the plan itself will involve a number of key players, including, in some cases, members of the steering committee. Most of the substantive work on implementing the mitigation programme takes place during execution—when the project manager will also monitor the effectiveness of the risk mitigation programme and seek to control its operation.

For example, take the threat "project manager leaves the company during the execution". Preemptives might include: put him/her under a long-term contract, offer a "golden handshake" (a completion bonus) or increase his/her remuneration. A contingency would be to have a deputy trained and ready to step in if required. A risk mitigation programme results in *changes* to the expected damage of certain threats and so the risk management process has to accommodate *two* values for each of likelihood, severity and expected damage. The first representing the values of these three variables *in the absence of* any mitigation and the second representing the values assumed *after* mitigation. These are identified as the pre- and post-values, respectively. The application of pre- and post-values for risk-related variables is discussed more fully in Sect. 6.4.1.

While the costs of risk mitigation must be taken into account in the project budget, the *actual* outlays will depend on whether or not the risk is realised. This is best explained by looking at two scenarios for each of the preemptives and contingencies:

Preemptives:

- Scenario A: preemptive cost is incurred immediately and remains insensitive to the threat's occurrence. For example, the leasing of an apartment for a project manager who lives overseas (to reduce the likelihood of his resignation).
- Scenario B: preemptive cost is only incurred if the threat IS NOT realised. For example, the offer of a completion bonus to a project manager.

Contingencies:

- Scenario A: contingency cost is incurred immediately and remains insensitive to the threat's occurrence. For example, the pretraining of a deputy to the project manager. This cost is incurred regardless of whether the project manager resigns or not.
- Scenario B: contingency cost is only incurred if the threat IS realised. For example, the costs of *inducting* the deputy into the project manager role. This cost is incurred only if the project manager DOES resign.

It is interesting to note that although, in general, contingency funds are necessary, some funding organisations do not allow them. In such situations, (undesirably) the contingency has to be "buried" by inflating other costs.

Preemptives may, in certain cases, increase the scope of the project through the addition of outputs beyond those implied by the ITO model. Consider the airport expansion project introduced in Chap. 5. As the project gets underway, one of the national security/intelligence services advises that terrorists have developed a new type of "wearable" explosive that can only be detected using advanced body scanners. In response, the funder decides to incorporate special scanning booths into the proposed security screening facility. Such a decision would increase the scope of the project (so that "advanced body scanners" are now in scope). This is yet another example of how different elements of a business case are interrelated (risk and scope in this case).

6.3.1.4 Mitigation Implementation

The risk management process continues throughout the project, as one of the mechanisms to support management of the project environment. A team member should be appointed as risk manager (a role that, by default, falls to the project manager or project administrator). The risk manager facilitates the identification of threats, maintains the risk register and oversees the planning/execution of the risk mitigation programme.

6.3.2 *The Risk Control Process*

The risk control process is concerned with ensuring that implemented programmes of risk mitigation remain effective throughout the project. The risk control process has three triggers:

A. **A threat already identified in the risk register is realised**. For example, the risk register includes the threat: "Process analysis consultant leaves" and we learn that he has just accepted an offer to join one of the company's competitors. When this happens, the risk manager and the person assigned responsibility for the risk should update the status of the entry (see the list of status flags under item #4 in Sect. 6.4.1 below) and consider:

    1. The appropriateness of any preemptives that were in place (for future reference).
    2. Steps that now need to be taken to activate any contingencies appearing in the risk register.
    3. A change of status to "Realised".

B. **A threat already identified in the risk register can now no longer be realised.** Using the previous example, the work of the analyst has been completed and he is no longer involved in the project. The risk manager now changes the status of the threat to "Closed".
C. **Regular reviews of the risk register**. At regular project review meetings, the person filling the role of risk manager would typically lead a discussion by selected project participants (particularly all those who have been assigned responsibility for a risk) to:

   - Decide if any of the details recorded for an existing threat (see next section) need updating.
   - Scan for any new threats and create an entry in the risk register.

6.4 Risk Management Tools

6.4.1 The Risk Register

The primary tool in risk management is the risk register, which is used to document the results of analysis and outline the mitigation programme being proposed for each threat. The risk register first appears in the business case and again later in the project plan. It is then updated regularly throughout project execution and outcomes realisation. Because it is a register, this tool takes the form of a table where rows are associated with instances (threats) and columns with attributes. A set of attributes making up a typical risk register would include columns for:

1. **Threat ID**. Typically, "T01", "T02" and so on.
2. **Threat title**. The triggering event expressed in "newspaper headline format".
3. **Description of threat**. The description includes details about the horizon assumed.
4. **Status**. An entry in the risk register develops overtime—as the risk management process unfolds. When first identified, only its name and description would normally be known, while by the end of the project it will be closed. This evolution is usefully summarised with a status flag drawn from the following sequence:

   - Identified: only a name and description of the threat are available.
   - Prepared: a mitigation programme has been proposed for the threat but not yet implemented.
   - Active: a mitigation programme for the threat has been implemented and is now being monitored.
   - Realised: the threat has occurred (and contingencies are now being exercised).
   - Closed: no further action is required, and the threat can now be ignored.

5. **Pre-likelihood**. Estimated for the threat in the absence of the proposed mitigating action—expressed as a value in the range [0,1].
6. **Consequences**. Description of any consequences (as a scenario) for the conduct of the project that would be attributable to the emergence of the threat. If the consequences are obvious, then this can be left empty.
7. **Damaging impact**. The damaging impacts on the project from the threat: expressed as a list drawn from the six generic forms of damage are summarised in Table 6.2.
8. **Pre-severity**. Estimated as the damage to the project (if the threat is materialised in the absence of the proposed mitigating action).
9. **Pre-expected damage**. Calculated as pre-likelihood * pre-severity.
10. **Risk mitigation actions**. Presented in the form of a set of preemptives and/or contingencies.
11. **The cost of the risk mitigation actions**. Cost is expressed as an indicative range.
12. **Post likelihood**. If preemptives are employed, this value will, in general, be less than the pre-likelihood.
13. **Post-severity**. If contingencies are employed this value will, in general, be less than the pre-severity.
14. **Post-expected damage**. This value will, in general, be less than the pre-expected damage. If this is not the case, then clearly the risk mitigation programme is at best having no effect, or, at worst, increasing expected damage. In either event, it would have to be rejected.
15. **The effectiveness of the proposed risk mitigation programme**. Gauged as "pre-expected damage" less "post-expected damage".
16. **Risk assigned to**. Identification of the person who has been given responsibility for implementing and monitoring the risk mitigation programme.

The reader can use this as the basis of a more comprehensive template that includes further information, such as date of last status change, date of realisation, actual severity of impact and so on.

Based on this analysis, a decision can be taken about which mitigating actions merit inclusion in the project's risk mitigation programme. Risk management demands effort and (possibly significant) additional expenditure, which the project's budget must accommodate.

The project manager holds the primary risk register, but (as was noted for the stakeholder register) there may be occasions when others hold "supplementary" registers. For example, the project owner may assemble a confidential risk register when premature release of information about a threat related to the project manager would cause unacceptable damage to the project.

In very large projects, risk management will be devolved out to teams, sub-teams, suppliers and contractors. Of the (possibly) numerous threats appearing throughout all these registers, only those meeting certain criteria are reported at the project level. The criteria used for such escalation are discussed below.

Consider the threat of a project manager quitting the organisation before project completion. This is an example of a generic threat—one that should appear in the risk register of every project. For ease of display here, the entry in Table 6.3 is arranged vertically. By way of contrast, a risk register template would typically have threats displayed as rows.

The decision to adopt the risk mitigation programme is then based on a trade-off by the project owner (or the funder) between the resulting increase the attractiveness of the project on one hand, and any increase in project cost arising from the risk mitigation programme on the other (see Sect. 7.3.3).

6.4.2 *The Risk Report*

Risk registers may have large numbers of entries—not all of which merit regular reporting and so it is desirable to highlight only selected risks for periodic review. Such an extract from the risk register is called a risk report.

A risk report is confined to those risks which meet *any* of the following criteria:

- Their pre-expected damage lies above some agreed threshold (whether they be existing or new threats).
- They have been realised since the last report.
- Their risk mitigation programme involves members of the steering committee.
- They have been closed since the last report.

As was the case with the stakeholder report, whenever a risk report is tabled, it should be accompanied by a link to the full register—thus allowing the readers who need more details to find it easily.

6.5 Issues and Their Management

6.5.1 *The Nature of Issues*

Consider the following scenario. An aluminium company has started work on a project to expand its bauxite mining operations in a remote area located 500 km by air from the nearest city. When the workplan was originally assembled, two situations prevailed:

- Equipment was to be moved between the city and the mine by either of two roads. One was 750 km long and crossed a river over a large bridge. The other had no bridges but was 1250 km long. The workplan assumed use of the shorter route.
- Only one company had planes suited to the relatively short runway available near the site. It offered one return flight each week to move workers between the site and the city.

Table 6.3 An example of an entry in the risk register

Threat attribute	Risk entry	Comment
Threat ID	T01	
Threat title	Project manager resigns	May or may not happen (note use of "newspaper headline format")
Description	Project manager (Fred Nurke) leaves the organisation over the next 2 months	A possibility based on some recent informal discussions with Fred
Status	Prepared	The programme will be discussed at the next steering committee meeting
Pre-likelihood	0.20	Because the project owner would be moderately surprised if the project manager left at short notice, pre-Li has been set to 0.20
Consequences	Negotiations with suppliers compromised	Caused by the loss of Fred's expertise and experience in this area
Damaging impacts	Benefits delayed Costs increased	
Pre-severity	$85,000	Assuming that the worth of the project in question is also gauged in dollars
Pre-expected damage	$17,000	The value ($17,000) is obtained as the product of pre-severity ($85,000) and pre-likelihood (0.20) (= $85,000 * 0.20)
Risk mitigation programme (RMP)	1. Preemptive: Offer a completion bonus 2. Contingency: Prepare a deputy to take over	Agreement to a completion bonus will reduce the likelihood of the project manager's departure Preparing a deputy to take over will reduce the damage (by shortening the time that the project would be without a manager)
Cost of threat mitigating actions.	$10,000	To deal with this particular threat, it is suggested to pay a bonus to the project manager. The bonus is set at $7,500. In addition, the costs of preparing and inducting the deputy amount to $2,500
Post-likelihood	0.10	The preemptive will halve the likelihood of the project manager leaving (from 0.20 to 0.10)
Post severity	$35,000	The effect of the contingency is expected to reduce the severity of the threat damage by $50,000 (from $85,000)
Post-expected damage	$3500	Together, the post-severity and post-likelihood result in a post-expected damage of $3,500 (= 0.10 * 35,000)
Effectiveness of RMP	$13,500	The combined changes in likelihood and severity result in a $13,500 movement of expected damage (from $17,000 to $3,500)
Risk assigned to	Project Owner	

Work was well advanced when two (unrelated) things happened: a historically unprecedented flood took out the road bridge and a new air transport company announced the introduction of a new 5 days/week service between the city and the mine. The loss of the bridge had not been identified as a threat, and so it was not entered into the project risk register.

Clearly, the implications of both of these developments demand exploration and consideration of some kind of response. The loss of the bridge and the availability of more frequent flights qualify as, what we call, project issues (Note that, unlike risks, issues need not have a damaging impact on the project). Here, both issues take the form of an event that has occurred—which then demand some sort of response by the project team. In a process somewhat akin to those employed for the management of stakeholders and risks, issues are: identified, analysed and managed through, in this case, an issues register.

An issue may need to be addressed with some sort of response. In the case of the two examples introduced above, responses might well take the forms of:

- Hiring of a barge and construction of wharves to ferry the equipment across the river.
- Rostering staff for shorter periods—allowing a reduction in the amount of on-site accommodation.

Issues can take any of three forms:

A. *Events* that have happened (regardless of whether their impact on the project is positive or negative): For example:

  - The curfew on the site has been reduced from 2100 through 0700 to 2200 through 0600.
  - On another project, the pumps we propose using have all experienced impellor failure.
  - A budget airline has just announced a new service to the city where one of our project sub-teams is based.

B. Important *questions* which, at this point, are unanswered: For example:

  - Is a reinforced ramp necessary for the 750-tonne crane?
  - Is stainless steel trim available for our control valves?
  - Can we move our regular meetings with the probity adviser from Fridays to Mondays?

C. Ad hoc *tasks* which do not appear elsewhere in project baseline documents (They are often workplan elements which have been overlooked). For example:

  - Fire safety inspectors are to confirm the safety of cladding before taking delivery.
  - The contract must be reviewed by our lawyers before signing.
  - No provision has been made for electrical services to the site.

Although, as pointed out earlier, a realised risk is also an event, this has no implications for its management—which is still guided by contingencies identified in the current *risk* mitigation programme.

The project environment is embedded in a continuous "foam" of issues—not all of which will be identified. Furthermore, not all of those identified merit a response.

6.5.2 Above-the-Line Versus Below-the-Line Issues

The bulk of issues arising in a project are below-the-line—that is they relate to the production, delivery or implementation of a project's outputs. The following are all examples of below-the-line issues:

- Steel has failed a tensile strength test.
- Delivery of raw materials has been advanced by four weeks.
- What is revised date for opening of the new rail line?
- The exchange rate has moved in our favour.
- Validate the As-Is models for our business process—using process simulation.

There will typically be a large number of below-the-line registers on a project —most of them distributed throughout various levels of the project team. For example, each subcontractor would be expected to maintain at least one. Most of the issues they contain are only of local interest, however, if their importance is judged as "critical", they will be escalated up through the project governance model for eventual mention in the next periodic status report (which is then tabled at the next steering committee meeting). The importance of an issue is discussed below.

Above-the-line issues relate to the planning and management of the project. The following are all above-the-line issues:

- The real estate agents have announced that our office lease will not be renewed.
- The funder has given priority to another project over ours.
- When will the funder approve the appointment of the project manager?
- The references of the selected applicant for the position of project administrator have yet to be checked.

6.5.3 The Issues Management Process

While the risk management process involves an analytical model, issues management is little more that clerical procedure (albeit a very important one). Issues management involves five subprocesses, as outlined in Table 6.4.

The issues management process is outlined in the following broad methodology:

1. **Issue identification**. Identify issues to which the project is now exposed. Issues will tend to be identified as they arise. Regularised scanning for them tends not

Table 6.4 The issues management process in outline

Subprocess	Outputs
Issue identification	A list of issues
Issue analysis	Details about each issue
Issue resolution formulation	Issues management programme
Issue resolution implementation	An implemented issues management programme
Issue monitoring and control	Updated issues management programme

to be terribly productive (because most of the events experienced during the project are irrelevant to its conduct). Project participants should be encouraged to recognise issues as they occur and get them into the relevant issues register as quickly as possible. Classification (as event, question or activity) is part of identification.

2. **Issue analysis**. This involves two aspects of the issue: importance and status.

   A. **Importance** relates to the need for a response by choosing an appropriate word from a simple wordscale such as:
   - Critical (must be escalated for inclusion in periodic project status reports)
   - High (demands a response)
   - Medium (a response is desirable)
   - Low (can be ignored).

   B. **Status** indicates where the issue sits in a stylised timeline. An entry in the issues register develops overtime—as the issues management process unfolds. When first identified only its name and description would normally be known, while by the end of the project, it will be closed. This evolution is usefully summarised with a status flag drawn from the following sequence:
   - Identified: only a name and description of the issue are available.
   - Parked: a decision has been taken to ignore the issue.
   - Prepared: a programme to resolve the issue has been proposed but not yet implemented.
   - Active: a programme to resolve the issue has been implemented and is now being monitored.
   - Resolved: no further action is required, and the issue can now be ignored.

3. **Issue resolution**. Formulate a response. This activity is completed in three steps: assign to a project participant to analyse, propose a course of responsive action and estimate the cost of the response (Typically, this will be someone filling a role in the steering, delivery or reference divisions of the project governance model).
4. **Issue resolution implementation**. If acceptable, implement the response.
5. **Issues monitoring and control**. The progress of the management programme is monitored and adjusted as required.

Issues management demands resources as additional input from the project team or as additional expenditure. As was the case with risks, these costs must be taken into account in the project budget.

6.5.4 Issues Management Tools

6.5.4.1 The Issues Register

The issues register has a structure somewhat like that of the risk register. The primary tool in issues management is the issues register, which is used to document the results of analysis and outline the response being proposed for each issue. The issues register first appears in the business case and again later in the project plan. Because it is a register, this tool takes the form of a table (see Table 6.5) where rows are associated with instances (issues) and columns with attributes.

A set of attributes making up a typical issues register would include columns for:

1. **Issue ID**. Typically, "I01", "I02" and so on.
2. **Issue name**. Expressed in "newspaper headline" format if it is an event, as a simple question or as an activity.
3. **Category**. Event, question or activity.
4. **Description**. One short sentence that describes the issue.
5. **Importance**. Declared as a value from the issues importance wordscale (typically: critical, high, medium, low—as discussed above).
6. **Status**. Declared as a value from the issues status wordscale (typically: identified, parked, prepared, active, resolved—as discussed above).

Table 6.5 The issues register

Issue attribute	Issue entry
Issue ID	I01
Issue name	Local landscape architect has decided to accept another engagement
Category	Event
Description	The firm did indicate that they were considering two engagements—but that they could resource only one of them
Importance	High
Status	Active
Response	Invite another landscape architect from a nearby town to submit a proposal
Cost of response	Negligible
Assigned to	Kim Dickinson, Procurement manager

7. **Response**. A response to the issue—typically in the form of a set of proposed actions.
8. **Cost of response**. The cost of the response to the issue—expressed as an indicative range.
9. **Issue assigned to**. Name of the person to whom issue has been assigned.

As was the case for stakeholders and risk, there is considerable flexibility with the incorporation of other columns—such as date of last status change.

Consider this example of a known event that has already happened, where one of the project's contractors will not resume the work as planned. For ease of display here, the entry is arranged vertically. By way of contrast, in the issues register template issues are displayed as rows.

As was noted for the risk register, the project manager holds the primary issues register, but there may be occasions when others hold "supplementary" registers. For example, the project assurance counsellor (when the role exists) may assemble a confidential issues register when premature release of information about an issue would cause irreparable damage.

Practicalities

Box 6.3: The Phenomenon of the "Exploding Issue Register"

The size of issue registers (gauged by the number of active entries) varies widely between projects. Size is influenced by a number of factors, including:

- The diligence (or zealotry) of the issues manager.
- The novelty of the project.
- The completeness of the description of the work to be undertaken during execution, for example as expressed by the number of project tasks and their size.

This last item bears further discussion. If the list of project tasks is incomplete or unreliable, then tasks that should have been identified remain hidden, until they can no longer be avoided and require immediate attention. The "sudden" emergence of such tasks makes them appear as issues and so the issue register rapidly expands with items that are really "missing project tasks".

If an issue register begins to grow rapidly, the project manager should investigate whether or not this has been caused by an incomplete or unreliable list of project tasks. If this is found to be the case, then clearly the project's workplan will need thorough revision.

6.5.4.2 The Issues Report

Issues registers may have large numbers of entries—not all of which merit regular reporting and so it is desirable to report only on those that demand the attention of others—especially the steering committee.

An issues report is confined to those issues which meet any of the following criteria:

- Their importance is "critical" or "high".
- Their response involves members of the steering committee.

As is the case with stakeholders and risk, whenever a stakeholder report is tabled, it should be accompanied by a link to the full register.

Summary

A project is constantly buffeted by influences that can push it off course. By classifying these influences as "risks" and "issues", we can bring to bear on the various tools that go a long way towards managing their impacts—especially any damaging impacts. Issues and risk management processes begin during initiation and continue throughout planning, execution and outcomes realisation. The risk and issues registers are effective management tools that can be tracked until project completion. The next chapter uses the concept of risk discussed here to decide on the attractiveness of projects to their funders.

Chapter 7
Project Attractiveness

Abstract When faced with opportunities to fund projects, organisations should address two questions: "Which candidate projects are suitable for funding?" And "How are they to be ranked in terms of their attractiveness for investment?" To answer these questions, it is necessary to assign each proposed project of some overall measure of "attractiveness for investment". The proposed *attractiveness* measure is a function of two variables: *expected return* and *riskiness*. Project return is analogous to the financial investment concept of Return on Investment (RoI), while riskiness is similar to investment risk. Unlike the situation in investment theory, however, the calculation of a specific value for return is, in general, extremely difficult because often projects also have non-monetary benefits and disbenefits. Despite this, because they can be systematically and consistently traded off against each other, the concepts of attractiveness, return and riskiness provide us with a useful conceptual framework for thinking about the valuation of projects. Every organisation will have a lower bound on attractiveness—below which a proposed project would not be funded.

7.1 Project Worth

When making a decision about approving project proposals, two key questions arise: "How are the candidate projects for funding to be ranked in terms of their attractiveness for investment?" And "Which of the ranked projects is suitable for funding?" To answer both questions, it is necessary to understand the "attractiveness for investment" of each proposal. If the level of attractiveness is below an organisational threshold, a proposal should not be funded. Based on financial portfolio selection models, the *attractiveness* measure discussed in this chapter is a function of two variables: *expected return* and *riskiness*. Project return (which is calculated from "worth") is analogous to the financial investment concept of (financial) Return on Investment (RoI), while riskiness is similar to investment risk.

In financial investment theory, assets are evaluated by trading off their return against their risk. Typically, this analysis is presented as a risk-return diagram, which is a cartesian plane with return displayed on a vertical axis and risk along a horizontal axis. When projects involve non-monetary benefits and disbenefits, it can prove

O. Zwikael and J. R. Smyrk, *Project Management*,
https://doi.org/10.1007/978-3-030-03174-9_7

difficult to use the tools and techniques of financial investment theory as they stand. Despite this, because they can be systematically and consistently traded off against each other, the concepts of attractiveness, return and risk provide us with an extremely useful conceptual framework for thinking about the valuation of projects.

The worth of a project is a function of benefits, disbenefits and costs:

$$\textbf{Worth} = \textbf{Fn (benefits, disbenefits, costs)} \tag{7.1}$$

Loosely, *worth "=" benefits—disbenefits—costs*, although this is rarely a correct expression because, in general, the components of benefits and disbenefits include incommensurate (non-monetary) units of measurement. While benefits are driven by target outcomes and disbenefits by undesirable outcomes, as explained in what follows, costs are driven by the resources consumed in the production of the project's committed outputs.

In general, each of the benefits/disbenefits/costs which make up Eq. 7.1 (both monetary and non-monetary) will take the form of either an inflow or outflow (over time) and, as a result, each could be displayed as a row in a table with columns related to time periods. When a *monetary-valued variable* flows over time, it can be "compressed' into a single figure which represents the value today of the entire future flow. This is achieved by *discounting* all future values and adding them together. The resulting sum is called the *present value* of the associated variable. When a present value has been calculated for all the *monetary-valued* variables in Eq. 7.1, some will be positive (inflows) and others negative (outflows). Adding all these together gives, what is called, a Net Present Value (NPV). Notionally, if NPV is positive, the project "generates more money than it consumes". Similarly, if NPV is negative, the project "consumes more money than it generates". If any of the variables represented by Eq. 7.1 are non-monetary, then the NPV calculated in this way is only *part of* a project's worth. If, however, all the variables represented by Eq. 7.1 are monetary, the worth of the project is *equivalent to* its NPV.

7.1.1 The Analysis of Project-Related Costs

When considering a prospective project for investment, the funder must take into account all monetary and non-monetary inflows and outflows attributable to the exercise. Monetary outflows are triggered by the acquisition (or deployment) of the resources demanded, not only by the work of the project itself, but also by the ongoing execution of the project's downstream processes. We split these monetary outflows into "project costs" and "operations costs". Project costs arise from the outlays to acquire the resources needed to conduct the exercise and consist of two types:

1. **Project execution costs** (otherwise known as "below-the-line" cost) arise from the acquisition of the resources required to produce the outputs identified in the project scoping statement. The overwhelming bulk of these costs are incurred during project execution.

2. **Project management costs** (otherwise known as "above-the-line" costs) arise from preparation for the project and its subsequent planning, management and administration.

Costs arising from both the execution and management of the project (together with all monetary-valued benefits and disbenefits) must be taken into account when calculating the project's NPV. Figure 7.1 shows how project and operations costs relate.

Four features of this diagram are worthy of note:

- Because it is a variant of the ITO model, there is a notional chronology flowing from left to right.
- The flow of project management costs (and hence total project costs) begins during planning.
- The period when project management costs are incurred overlaps part of the operational environment. This arises from the fourth of a project's global phases (outcomes realisation) in which the early part of the operational environment is also treated as part of the project environment.

We now address the question "How are the costs of risk mitigation to be handled?" When considering a proposed risk mitigation programme, it is necessary to compare two project scenarios: one in which the proposed risk mitigation programme has been accepted and the other in which it has been rejected. Under the first scenario, the cost of risk mitigation is included in *execution costs*, while in the second it is not included.

Some of the resources demanded by a project involve outlays of cash (to purchase products and services), while others involve the deployment of existing resources (such as operations staff who have been seconded to the project). In this latter case,

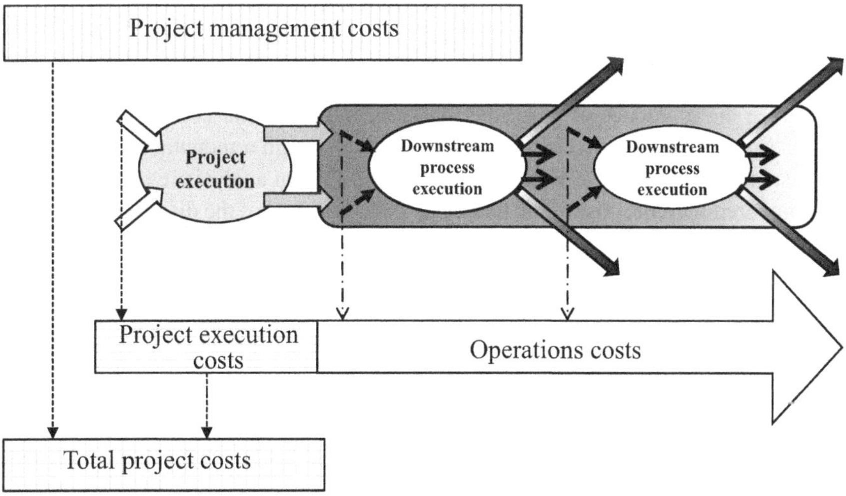

Fig. 7.1 Project costs elements in relation to the ITO model of a project

the labour involved must be given a monetary value. To obtain this, one should then ask, "What is the net financial impact on the organisation of the deployment?" If, for example, contractors must be hired to take over the operations roles that have been left vacant by the deployment, then the assigned staff is costed at the remuneration rates agreed for the *contractors*. When there is no such "backfilling", it is common practice to base costs on salaries (possibly adjusted for on-costs such as holidays).

7.1.2 Project Worth: Incommensurate Units of Measurement

The measures associated with all of a project's worth variables are usually incompatible, making it difficult in practice to calculate an overall numeric value for worth. This now raises the question "Is it ever feasible to measure worth with a single variable?" As is clear from the vaccine example introduced in Box 7.2 that worth can be calculated simply and unambiguously if all benefits, disbenefits and costs can be given a meaningful monetary measure, but this is a relatively rare situation.

Numerous approaches have been proposed to address this problem, including, for example, economic cost–benefit analysis, multiple-criteria analysis and making trade-offs amongst the variables. When considering the worth of a project to the funder, it is, in general, not helpful to convert all non-monetary-valued variables into money flows. Multi-criteria approaches, such as the Multi-Attribute Utility/Value Theories (MAUT/MAVT), data envelopment analysis (DEA) and Analytical Hierarchical Process (AHP) can enable a more comprehensive consideration of multiple aspects (Chih and Zwikael 2015), however the time and resource constraints imposed on most projects by their business cases make use of such techniques impractical.

The practice of assigning arbitrary monetary value (such as dollars) to any non-monetary-valued target outcomes bears further discussion because it sometimes complicates the analysis of project worth rather than simplifying it:

1. **Equivalence**. A set of monetary-valued effects is not *equivalent to* the non-monetary-valued variable from which they derive. Take the case of a project to reduce the incidence of domestic violence in a particular community by 1,000 cases/year. Assume that it has been decided to assign a monetary value to each case using some sort of economic model and that an analysis reveals a shortfall compared with the cost of the initiative. Is that the end of the discussion, or could a case still be mounted in support of the initiative? If we allow a trade-off between *the way the funder values* the reduced number of domestic violence cases and project cost, then a decision could very well be taken in favour of funding, despite the apparent negative monetary worth.
2. **Conversion**. The application of a monetary conversion factor does not convert a non-monetary variable into a real flow of money for the funder (Whether or not

a project *should* generate financial inflows for the funder is an entirely separate issue).

3. **Measurement**. As a gauge of eventual success, measurement of a monetary surrogate is not equivalent to measuring the target outcome. Using the example of the domestic violence initiative, take the situation where it has been decided to give each incidence of domestic violence a notional dollar value and, accordingly, to set a target outcome of "reduced costs of domestic violence". What variable will be measured to judge success (ex post)? There appear to be only two options: the actual number of incidents of domestic violence or the actual value of the monetary surrogate. If, however, the latter approach is used, how is the surrogate to be calculated? Since it is obtained as the product of the count of cases and the assumed monetary value of each case, then success will be sensitive only to the count of cases, the monetary value of each case can have no bearing on the issue.
4. **Pooling**. The fourth point about the limitations of monetary-valued variables requires some preliminary discussion of the concept of "pooling". Consider a project in which some of the benefits and some of the disbenefits happen to take the form of "natural" flows of money (Note that in general, the *costs* of all projects can be represented as money flows). Depending on the particular circumstances, various subsets of these flows can be pooled by particular stakeholders. For example, an initiative to boost tourist visits to a historic town is to be funded jointly by a state tourism authority and local businesses with the target outcome of "increased sales by local businesses". A variety of scenarios can be postulated for this project, each based on different patterns of pooling the money flows that it generates, and each giving rise to its own peculiar treatment of the worth of the venture. Compare these two scenarios:

   - "Y" (in which each firm looks after its own sales, costs and project levies without any sharing of any moneys).
   - "Z" (in which all money from all sales, costs and levies on all firms flow through a single bank account administered by the state tourism authority and eventually disbursed according to some agreed formula).

   Clearly, the valuation of the project by each of the participants (and their resulting behaviours) will be heavily influenced by the existence of a pool (as in scenario "Z") or otherwise (as in scenario "Y").
5. **Valuation**. Regardless of whether an outcome is measured in monetary or non-monetary terms, its benefit must be gauged from the funder's perspective, not the beneficiary's. Consider a project funded by a city council to attract a major international powerboat race, with target outcomes of increased visitations by tourists to the city and increased volumes of sales by local tourist operators. Assume that the project generates significant revenue for an airline that operates scheduled flights to the city, but which is based elsewhere and does no significant business locally. While the outcome of increased contribution margin (gross profit) to the airline certainly has a monetary value, in terms of benefits to the funding council, this outcome might well have zero value.

6. **Empirical validity**. If a non-monetary-valued target outcome is given a monetary value by applying a formula, the coefficients in that formula must be capable of validation, whereby their value can be tested empirically as correct or incorrect. Take the scenario in which, for the domestic violence initiative, a monetary value has been assigned to each case using a formula derived from an economic model. Because, as pointed out in 3 above, the financial valuation of the project is sensitive to both the number of cases and the value assigned to each, the question arises "How do we know if the monetary value being used is correct?" Clearly if it is incorrect, then the calculated financial worth of the project will be wrong. If an assigned or assumed value in a conversion formula cannot be empirically validated, then the formula cannot be used meaningfully in the analysis of the project.

7.1.3 Accommodating Both Monetary and Non-monetary Worth Values

While project costs can always be expressed in monetary units, typically, the other two determinants of project worth (benefits and disbenefits) will have a mixture of monetary and non-monetary units of measurement. In such cases, it is possible (and desirable) to simplify the (usually disparate) collections of variables that represent a project's benefits, disbenefits and costs. This can be done when two conditions are met: that the "natural" unit of measure for particular variables is dollars (in other words some variables take the form of a real flow of money) and that the funder is able to pool all these flows. In those cases, all the monetary-valued variables (including relevant benefits, disbenefits and costs can be "compressed" into an NPV). When NPV has been calculated in this way, the only terms that now remain are the non-monetary-valued components of benefits and disbenefits. To determine a project's worth, the NPV of the project must be traded off against all these non-monetary-valued variables—as suggested in the example in Box 7.1. In such cases, the worth of a project can be expressed as

$$\textbf{Worth} = \textbf{Fn}\ (\textbf{Non-monetary-valued-benefits}, \textbf{Non-monetary-valued-disbenefits}, \textbf{NPV}) \tag{7.2}$$

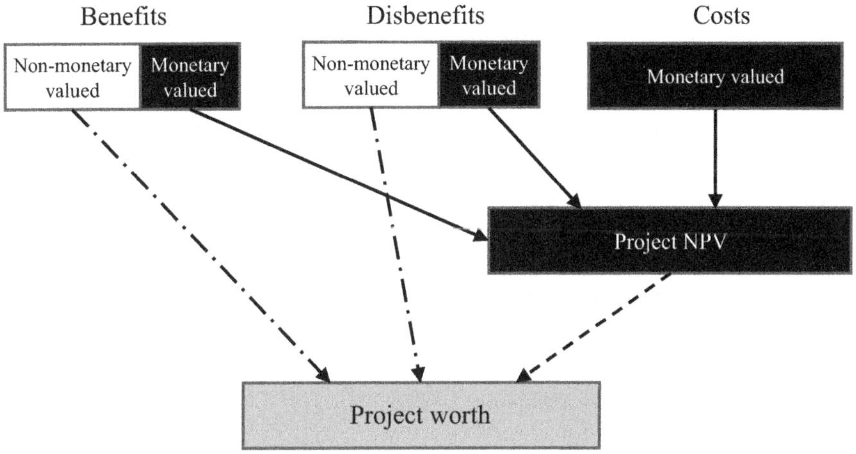

Fig. 7.2 Determination of project worth using a combination of monetary and non-monetary values

Figure 7.2 illustrates the conceptual elements of this function.

Illustrative example

Box 7.1: Trading-off Net Present Value (NPV) against Non-monetary-Valued Target Outcomes

Consider a project by a construction company to institute a major industrial safety campaign. Currently, the safety practices that the organisation currently has in place are extremely poor, resulting in an unenviable accident record and high insurance premiums. The company also has (what is in the eyes of the Board) an unacceptable ranking in an annual award for excellence conducted by the professional association of builders to which the firm belongs.

Two outcomes are sought from the initiative: reduced insurance costs and increased ranking in the excellence awards. The first has been specified as an ongoing annual reduction of \$250,000 and the second as a move in industry rankings, from the 90th percentile to the 10th percentile. The project will involve an initial outlay of \$1 M with ongoing costs of \$185,000/year. Because the three dollar-valued variables (project cost and reduced insurance premiums) will be automatically pooled by the construction company, it is meaningful to combine them into an NPV. Assuming that reduction in insurance premiums and that ongoing costs will flow in perpetuity, at a 10% discount rate (the assumed market cost of money), the project has an NPV of −\$350,000.

This figure has been obtained in the following steps:

- Calculate the annual cash flow (65,000 = 250,000–185,000).
- Discount this flow at 10% (650,000 = 65,000/10%).
- Subtract the upfront cost of the initiative (−350,000 = 650,000 − 1,000,000)

In other words, the net financial gain from the venture falls short of the project's costs, so that the initiative has a negative financial value of $350,000. The *worth* of the project can now be decided by trading off the "loss" of $350,000 against the firm's improved ranking on the industry ladder. The project funder will have to make this subjective judgement. The *attractiveness* of the exercise can then be decided when its riskiness is eventually determined from its risk register.

7.1.4 Evaluating a Project's Worth and Return

The worth of a project attempts to account for desirable and undesirable outcomes, as well as the notional costs of *all resources consumed through until the end of the project*. When costs are first estimated (during initiation), they cover all the resources required not only during planning and execution but also all those that are necessary to service the entire future operational environment (within which the downstream processes are executed). In what follows, we assume that worth has only dollar-valued components.

The worth of a project has a number of variants:

1. **Initial project worth** is calculated using the raw target values for benefits, disbenefits and costs without any provision for either expected damage or the costs of risk mitigation (discussed in Sect. 7.2.3), i.e. before a Risk Mitigation Programme (RMP) is implemented (see Eq. 7.3).
2. **Expected project worth (in the absence of any risk mitigation)**, is derived from initial project worth, while also taking into account the expected damage from the various identified threats (but before accounting for the costs of any risk mitigation) (see Eq. 7.3).
3. **Expected project worth (following risk mitigation)**, is derived from the initial worth and expected damage (following implementation of a risk mitigation programme) (see Eq. 7.4).
4. **Achieved worth** which is based on the realised values for benefits, disbenefits and costs as it is measured at the end of the project.

These variants are expressed in the following equations:

$$\begin{aligned}\textbf{Initial\_project\_worth} = \textbf{Total\_benefits} - \textbf{Total\_disbenefits} \\ - \textbf{Total\_project\_costs}\end{aligned} \tag{7.3}$$

$$\textbf{Expected\_worth\_pre\_rmp} = \textbf{Initial\_project\_worth} - \textbf{Expected\_damage\_ pre\_rmp} \tag{7.4}$$

$$\begin{aligned}&\textbf{Expected\_worth\_ post\_rmp}\\&= \textbf{Initial\_project\_worth} - \textbf{Expected\_damage\_ post\_rmp}\\&\quad - \textbf{Rmp\_costs}\end{aligned} \tag{7.5}$$

$$\textbf{Achieved\_Worth} = \textbf{Realised\_Benefits} - \textbf{Realised\_Disbenefits} - \textbf{Realised\_Costs}) \tag{7.6}$$

Illustrative example

Box 7.2: A Project to Manufacture a Genomic Vaccine
A large pharmaceutical company ("LPC") has approved a project to manufacture a genomic vaccine for Ebola ("GVE")—to be offered as an alternative to an existing conventional product. The new product has already been approved for sale by the US Food and Drug Administration. It is expected that the new product will cannibalise some of the markets for an existing product—leading to reduced production volumes.

The worth of the initiative can be derived from the following parameters (all of which are represented here by their *present values* over an expected product life of five years). Note that in this particular case, both benefits and disbenefits have dollar values. In what follows, the name of the variables used in the equations throughout this chapter is also shown (in italics) for each variable:

1. **Benefits**:
   - New sales of the new drug = $M2,000 *(Target_sales).*
   - Decreased costs of manufacturing the existing product = $M15 *(Reduction_manufacturing_costs_existing_product).*
   - **Total benefits = $M2,015** (2,000 + 15) *(Total_benefits)*

2. **Disbenefits**:
   - Decreased revenue from cannibalisation of the existing product = $M150 *(Lost_Revenue_Cannibalisation_Existing_Product)*
   - New product manufacturing costs = $M140 *(New_product_manufacturing_costs)*
   - **Total disbenefits = $M290** (150 + 140) *(Total_disbenefits)*

3. **Costs:**
   - **Total project costs (in the absence of any risk mitigation) = $M540** *(Total_project_costs_pre-rmp)*
4. **Worth** (= Benefits – Disbenefits – Costs):
   - **Initial project worth =$M1,185** (2,015 – 290 – 540)

A comment about the appearance in Box 7.2 of manufacturing costs for the new drug as a *disbenefit* rather than a *cost* is necessary. Intuition suggests that, as a "cost", this should be included in the cost term. While arithmetically both treatments would yield a worth of $M1,185, it is important to understand the logic behind the categorisation. In the IPO model, the work of producing the outputs from a process (any process) requires certain resources. The cost of producing the outputs from the process is defined as the value of those resources consumed by the process. In the example of the new drug, two distinct processes are involved in the initiative:

- Project execution—the outputs from which would include the new factory, a marketing campaign, enabling information systems/infrastructure, and a new sales unit.
- A downstream operational (manufacturing) process—the output from which is the drug itself (in various packages).

Is execution of the downstream operational process necessary to produce the *project's outputs*? Clearly no, because the product cannot be manufactured until *after* the factory is commissioned. We can conclude, therefore, that future operations costs are not components of project costs (the fact that the *project's target outcomes* cannot be generated without executing the operational process has no bearing on the costs of *the project's outputs*). At the same time, however, when projects result in operations costs changing (up or down), such changes must be accounted for in the equation of worth. This is achieved by simply classifying the *change* in operations costs as either a benefit (if the change is favourable) or a disbenefit (if the change is unfavourable).

Continuing the example in Box 7.2, the decision by LPC to invest in the GVE project requires (amongst other things) an estimate of the project's expected return. This is obtained from the estimates of worth, expected damage (arising from the risks to which the exercise is exposed) and project cost.

Assume in the case that the pre-expected damage from the GVE risk register (in the absence of any risk mitigation) is $M134, but after mitigation (cost of risk mitigation is $M13), its post-expected damage is $M63. Therefore,

Expected_worth_ pre_rmp = *Initial_project_worth – Expected_damage_ pre_rmp = $M(1,185 – 134) = $M1,051*

Expected_worth_ Post_rmp = *Initial_project_worth – Expected_damage_ post_rmp – rmp_costs = $M(1,185 – 63 – 13) = $M1,109*

It should be noted that, because the change in expected damage can be greater than or less than the costs of mitigation, post-expected worth can be above or below pre-expected worth.

Investment decision is commonly based on Return on Investment (RoI) rather than NPV—where RoI = NPV/project costs (RoI is conventionally measured as a %). Just as we use the term "worth" instead of NPV when the variables in Eq. 7.1 are incommensurate, so we use "return" instead of RoI. More is said about this in the discussion of project portfolios in Sect. 7.4). Return is a generalised version of RoI and is conceptually derived as

$$\mathbf{Return} = \mathbf{Worth/Total\_Project\_Costs} \tag{7.7}$$

After taking the costs of risk mitigation into account, the total project cost is \$M553 (= \$M540 + \$M13). We can now calculate the following return values:

> ***Expected return (based on expected project worth in the absence of any risk mitigation)***
> *Return-pre_expected_worth = Expected_worth_pre_rmp/Total_project_costs_pre-rmp = 1,051/540 = 195%.*
>
> ***Expected return (based on expected project worth following risk mitigation)***
> *Return_post_expected_worth = Expected_worth_post-rmp/Total_project_costs_post_rmp = 1,109/553 = 200%.*

This means that if the risk mitigation programme is implemented, every one dollar invested in the project, is expected to generate \$2.00.

From the literature

Box 7.3: The Concept of Satisficing

In his seminal work "Models of Man", Simon (1976) examined the situation encountered by decision makers who face high levels of complexity in choosing a course of action. He noted that decisions are made within the constraints of available information and the analytical capacity of the decision maker.

He labelled this constrained approach to decisions *bounded rationality*. In order to make a defendable choice, he suggested considering only those options that lead to a "satisfactory" result (A "satisfactory" result being one that meets a declared threshold). This method of simplification he called *satisficing*—to distinguish it from *optimising*—in which the "best" decision is sought.

The assumption of satisficing behaviour has implications for our treatment of attractiveness (this chapter) and success (Chap. 8).

7.2 Project Riskiness

7.2.1 The Statistical Distribution of a Threat's Damage

Riskiness is a measure of the uncertainty that the worth of a project will achieve a particular level. In other words, riskiness is a measure of the possible spread of results that might be experienced. A wide spread of possible values of worth makes any particular target range less certain that would be the case with a narrow spread of the possible values of worth. This section provides the statistical foundations to explain how the *distribution of damage from the threats to which a project is exposed* can be used to derive the *riskiness of a project.*

The discussion in Chap. 6 of likelihood (Li), severity (Se) and expected damage (ED) implies that each threat to a project is, what statisticians call, a (non-repeated) Bernoulli trial. A Bernoulli trial is an "experiment" in which there are exactly two results (threat realised, threat not realised) and the probability of the threat being realised is known in advance.

As discussed in Chap. 6, the metric for likelihood is our *judgement* of the probability of the threat being realised, and so is a number in the interval [0, 1]. The metric for the severity of the damage arising from a threat is expressed as the amount of the project's target worth that would be lost if the threat arises.

Each threat therefore has a probability distribution—formally called a *Bernoulli distribution.* In the probability distribution diagram (Fig. 7.3), the x-axis shows the amounts of damage related to a threat (in the same notional units adopted to measure the project's worth), while the y-axis shows the probability of that damage being incurred [0, 1].

For each project threat, a Bernoulli distribution has two spikes: one located at zero on the x-axis (corresponding to the result "Threat not realised") and another

Fig. 7.3 An example of a Bernoulli distribution for a threat

with an x-value equal to the damage associated with the result "Threat realised"). A Bernoulli distribution has a *mean* (which always equals the expected damage from the threat and a *standard deviation* (indicating the "spread" of the two spikes). The mean of a Bernoulli distribution is calculated as the *probability of the event occurring* multiplied *by the severity of the consequent damage*, as is indicated in Eq. 7.8.

$$\textbf{Expected\_damage} = \textbf{Severity\_of\_damage} * \textbf{Likelihood} \tag{7.8}$$

Figure 7.3 illustrates a Bernoulli distribution for a threat with a likelihood of 0.2—for which the resulting damage is $25—and hence an expected damage of $5 (=0.20 * 25)—in other words, the expected damage from the threat is $5.00.

We now turn our attention to the way in which the damage associated with the entire risk register is "spread". The following treatment of this spread gives rise to a formal measure of a project's riskiness.

7.2.2 *The Uncertainty of Damage from a Project Threat*

If we were given the opportunity of randomly re-running the history of a particular threat to a project, say, 100 times, what sorts of results might we observe? We could find that on every run the threat was found to occur. Clearly, the probability of such a result is extremely small. The same is true of a result in which the threat did not occur in any of the 100 reruns. Other runs would give rise to randomly varying numbers of occurrences of the threat in question.

We now consider the question "How *spread* would be all the possible values of damage from this threat over our 100 trials? Statisticians measure such spread using *variance* and *standard deviation* (Standard deviation being the square root of variance). To calculate the variance of our Bernoulli distribution we use the same three parameters appearing in the calculation of the mean (likelihood, 1-likelihood, severity) but apply them to a different formula. Threat variance is calculated as *the likelihood of the threat occurring* multiplied by *the likelihood of it not occurring* multiplied by *the square of the severity* (Eq. 7.9). In the example used for Fig. 7.3, the variance would be 100 (= 0.20 * 0.80 * 25 * 25). Using Eq. 7.10, the standard deviation is then found as 10 (=the square root of 100).

$$\textbf{Threat\_damage\_variance} = \textbf{Likelihood} * (\textbf{1} - \textbf{Likelihood}) * (\textbf{Severity})^2 \tag{7.9}$$

$$\textbf{Threat\_damage\_standard\_deviation} = \sqrt{\textbf{Threat\_\_damage\_variance}} \tag{7.10}$$

An example of the use of these equations is provided in Table 7.1 where:

- The entries for Threat A are as discussed above.

- For threat B, the variance is calculated as 20.25 (=0.10 * 0.9 * 15 * 15) and the threat standard deviation as its square root (=4.50).

Recall that the expected damage of a threat is a measure of the amount of damage that we expect to occur, while the standard deviation of the damage from a threat is a measure of our *uncertainty* that we will experience that particular amount of damage. A large standard deviation means that we could well finish up with damage that is very different from what we expect, whereas a small standard deviation means that we are going to experience a level of damage that is much closer to what we expect. The relationship between likelihood, expected threat damage and the standard deviation of threat damage bears further discussion because it is counterintuitive. Consider threat A for which the anticipated resultant damage (in the absence of any mitigating actions) is $25. Using the example shown in Fig. 7.3, Table 7.2 shows what happens to the expected damage, variance and the standard deviation of damage as likelihood (alone) changes.

Some observations on Table 7.2:

- **As likelihood rises so does expected damage**: As the likelihood of a threat increases from 0.1 to 0.9, the expected damage also rises from 2.5 to 22.5.
- **As the Likelihood of a threat increases, the standard deviation of the threat damage initially rises, but then falls**: The table reveals that as likelihood rises the standard deviation of the threat damage increases to a maximum of 12.5 (at

Table 7.1 Calculating threat standard deviation

Threat #	Severity ($)	Likelihood	Expected damage ($)	Threat damage variance ($^{2})	Threat damage standard deviation ($)
A	25	0.2	5.00	100.00	10.00
B	15	0.1	1.5	20.25	4.50

Table 7.2 The response of expected damage and the standard deviation of damage to changes in likelihood

Severity ($)	Likelihood	Expected damage ($)	Threat damage variance ($^{2})	Threat damage standard deviation ($)
25.00	0.1	2.50	56.25	7.50
25.00	0.2	5.00	100.00	10.00
25.00	0.3	7.50	131.25	11.46
25.00	0.4	10.00	150.00	12.25
25.00	**0.5**	**12.50**	**156.25**	**12.50**
25.00	0.6	15.00	150.00	12.25
25.00	0.7	17.50	131.25	11.46
25.00	0.8	20.00	100.00	10.00
25.00	0.9	22.50	56.25	7.50

likelihood = 0.5), and then gradually falls back. Because this is counter-intuitive, some explanation is warranted. Consider the threat analysed in Table 7.2. If we gauge that its likelihood as zero, then we are declaring that it **cannot occur**, implying that its expected damage is zero and so the project's expected worth will be equal to its target worth (with zero uncertainty). If, however, we decided that its likelihood is one, then it **must occur**, implying that the project's expected worth will be equal to its target worth minus the damage from the threat (also with zero uncertainty) (The reader will note that expected damage and standard deviation both achieve the same a maximum (of 12.5) at likelihood = 0.5. This is a completely general observation that is true for all such distributions).

7.2.3 Calculating a Project's Riskiness

We have introduced the use of standard deviation as a measure of the "uncertainty" surrounding the realisation of a **specific** *threat*. We now need to consider the uncertainty associated with a **project's** *expected worth* (and hence its anticipated return). Consider Eqs. 7.3 and 7.4. The expected worth of the project will have the same standard deviation ("variability") as the damage represented by the *entire risk register* (where all the threats to a project are catalogued). We then define project riskiness in terms of both total project costs and the standard deviation *of the damage* associated with all threats in the risk register.

The following analysis assumes that all threats in the risk register are statistically independent. This means that the likelihood and severity of every threat in the risk register remains unchanged in response to the realisation of any other threat. Given this assumption, the *expected value of overall (project-level) damage* and *variance of overall (project-level) damage* can be found by simply summing the corresponding parameters for all of the individual threats recognised across the entire risk register The statistical proof of these relationships lies beyond the scope of this book.

$$\mathbf{Project\_expected\_damage} \\ = \mathbf{Sum\ over\ the\ entire\ risk\ register\ (Threat\_expected\_damage)} \quad (7.11)$$

$$\mathbf{Project\_damage\_variance} \\ = \mathbf{Sum\ over\ the\ entire\ risk\ register\ (Threat\_damage\_variance)} \quad (7.12)$$

$$\mathbf{Project\_damage\_standard\_deviation} = \sqrt{\mathbf{Project\_damage\_variance}} \quad (7.13)$$

$$\mathbf{Project\_riskiness} = \mathbf{Project\_damage\_standard\_deviation/Total\_project\ costs} \quad (7.14)$$

There are two values of riskiness that are of interest to us—measured respectively before and after the risk mitigation programme is in place:

$$\textbf{Riskiness\_pre\_rmp} = \textbf{Project\_damage\_standard\_deviation\_pre\_rmp} \\ /\textbf{Total\_project\_costs\_pre} - \textbf{rmp} \tag{7.15}$$

$$\textbf{Riskiness\_post\_rmp} = \textbf{Project\_damage\_standard\_deviation\_pre\_rmp} \\ /\textbf{Total\_project\_costs\_pre} - \textbf{rmp} \tag{7.16}$$

So how can the risk register play out in the course of a project? At one extreme, we could find that when the project concludes, no threats have been realised. At the other extreme, we could find that every threat has been realised (both situations would have extremely small probabilities and are rarely going to be experienced in real life). Clearly, any combination of threats from the risk register could occur. If, for example, there are 50 threats in the risk register, there are 2^{50} possible different combinations of threats that might be experienced (this is an extremely large number with 16 digits).

In general, we can show that, as the number of threats increases, the probability distribution of damage will tend to have:

- Shorter spikes within a bell-shaped envelope.
- More numerous spikes (for N threats, there could be *up to* 2^N spikes).
- A larger mean (with the envelope shifted towards the right on the x-axis).
- A wider spread of values along the x-axis.

The standard deviation offers a robust measure of this spread—which we call *riskiness*. Chapter 6 discussed the need for two measurements (pre and post) for each of likelihood, severity and expected damage associated with a threat. While the overall *riskiness* of a project will, therefore have a pre- and post-value, it is only the post-riskiness that is of concern to the funder, because the funding decision will be based on the project *as proposed* (which takes into account all of the implications of risk mitigation).

Tables 7.3 and 7.4 together provide a numerical summary of a hypothetical risk register for the genomic vaccine illustration (for which, for ease of discussion, worth is measured in dollars). The extract highlights post-values for various parameters (assuming that an RMP has been adopted). Expected damage is calculated for each threat based on Eq. 7.8, threat variance based on Eq. 7.9 and threat standard deviation based on Eq. 7.10. To obtain the corresponding parameters for the project overall, column totals are calculated for: severity, expected damage (based on Eq. 7.11) and variance (based on Eq. 7.12). Project riskiness is derived, not as a column total, but (from Eq. 7.13), as the square root of the project damage variance.

The expected project worth (following risk mitigation) can be found using Eq. 7.5:

$$\textbf{Expected\_worth\_post\_rmp} = \textbf{Initial\_project\_worth} \\ - \textbf{Expected\_damage\_post\_rmp} - \textbf{Rmp\_costs} = \text{M\$1,109}\ (= 1{,}185 - 63 - 13).$$

Table 7.3 Calculating pre-mitigation riskiness for the genomic vaccine example

Threat #	Severity ($000)	Likelihood	Expected damage ($000)	Threat damage variance ($^{2}000000)	Threat damage standard deviation ($000)	Cost of risk mitigation programme ($000)
A	5,000	0.10	500	2,250,000	1,500	1,500
B	26,000	0.10	2,600	60,840,000	7,800	700
C	800	0.70	560	134,400	367	100
D	6,400	0.60	3,840	9,830,400	3,135	2,500
E	7,500	0.20	1,500	9,000,000	3,000	300
F	500,000	0.25	125,000	46,875,000,000	216,506	4,000
G	8,400	0.05	420	3,351,600	1,831	3,500
H	650	0.01	7	4,183	65	400
Totals	554,750		134,427	46,960,410,583	216,704	13,000

Table 7.4 Calculating post-mitigation riskiness for the genomic vaccine example

Threat #	Severity ($000)	Likelihood	Expected damage ($000)	Threat damage variance ($^{2}000000)	Threat damage standard deviation ($000)
A	2,000	0.08	160	294,400	543
B	12,000	0.09	1,080	11,793,600	3,434
C	200	0.50	100	10,000	100
D	3,400	0.50	1,700	2,890,000	1,700
E	2,500	0.15	375	796,875	893
F	300,000	0.20	60,000	14,400,000,000	120,000
G	850	0.01	9	7,153	85
H	70	–	–	–	–
Totals	321,020		63,424	14,415,792,028	120,066

By way of contrast, the expected project worth (following risk mitigation) can be found using Eq. 7.4:

$$\begin{aligned}&\textbf{Expected\_worth\_pre\_rmp} = \textbf{Initial\_project\_worth}\\&\quad - \textbf{Expected\_damage\_ pre\_rmp} = \text{M\$1,051}(= 1{,}185 - 134).\end{aligned}$$

This means that the risk mitigation programme increased the expected worth of the project by M$58 (1,109 – 1,051). It should be noted that a risk management programme which happens to reduce a project's worth may still be adopted if, at the same time, it increases the project's attractiveness by an acceptable amount.

In addition to its worth, the other determinant of project attractiveness is its riskiness. The genomic vaccine project's riskiness (post-RMP) is M$120, which is lower than the (pre-RMP) value of M$216 (before introducing the risk management programme). Riskiness can also be calculated as a percentage (to allow easy comparisons among different projects):

Riskiness (in the absence of any risk mitigation)

Riskiness_pre_rmp = Standard_deviation_pre_rmp

= Standard_deviation_pre_rmp/Total_project_costs_pre-rmp = 216/540 = 40%.

Riskiness_post_rmp = Standard_deviation_post_rmp

= Standard_deviation_post_rmp/Total_project_costs_post_rmp = 120/553 = 22%.

7.3 The Project Attractiveness Map

7.3.1 The Dimensions of the Project Attractiveness Map

As discussed in Sect. 7.1, the finance discipline uses the risk-return diagram to decide on an optimal portfolio of assets given multiple potential investment opportunities. Risk aversion (which is implied by our assumption of satisficing behaviour on the part of a project funder) is characterised by the attachment of decreasing levels of value to additional amounts of return on an investment. In other words, each additional dollar of return is valued less highly than the previous dollar. Investment by a risk-averse organisation in high-risk projects demands therefore, that returns be correspondingly higher. In *project* investment decisions, we use the attractiveness map. Figure 7.4 shows a plane defined by the two variables on which a project's attractiveness is based—return and riskiness. Where is a *proposed* project to be located on this plane? The y-axis value is calculated as *expected* return, while the x-axis value (riskiness) is obtained by calculating the standard deviation of the returns distribution generated by the entire risk register. Because the focus of our attention is *proposed* (as distinct from *completed*) projects, most of the following discussion concerns *expected* (as distinct from *achieved*) worth.

The finance discipline uses the risk-return diagram to decide on an optimal portfolio of assets given multiple potential investment opportunities. Similarly, in project investment decisions, we use the attractiveness map. Figure 7.4 shows a plane defined by the two variables on which a project's attractiveness is based—return and riskiness.

As discussed in Sect. 6.2.1, the location of a *completed* project (for which almost all risk has effectively evaporated) will have an x-value close to zero (over near the y-axis), and a y-value obtained as the *actual* worth generated by the (now completed) project.

The plane is divided into two regions separated by a curve identified as the "project investment frontier". The project investment frontier is a plot of all combinations of expected worth and riskiness that a funder would be prepared to accept as a worst case. Those points that lie above (or to the left) of this frontier represent projects that are attractive for funding, while those below (or to the right) are unattractive. The project investment frontier (which is unique for each organisation or decision maker), is based on the funder's preferences and external constraints. The location of the project investment frontier can also change over time. The points "A" and "U" show two proposed projects.

Some further comments about the diagram are the following:

- The return (y) axis could conceivably range from positive down through negative values, although any project with a return of less than 0% would never be acceptable for funding.
- The riskiness (x) axis can only take on positive values because it measures the riskiness of a project in terms of the standard deviation of the distribution of returns implied by all threats identified in the risk register (standard deviation can never be negative).
- There will be a certain minimum level of anticipated return for a zero-risk project below which a project will not be funded (shown as project "N"). This accounts for initiatives that are so low in anticipated return that they do not even merit the time and effort of processing a business case.
- Attractiveness increases as a project moves up and/or to the left.
- Note how "U" is less attractive than "A" (despite having a higher return).

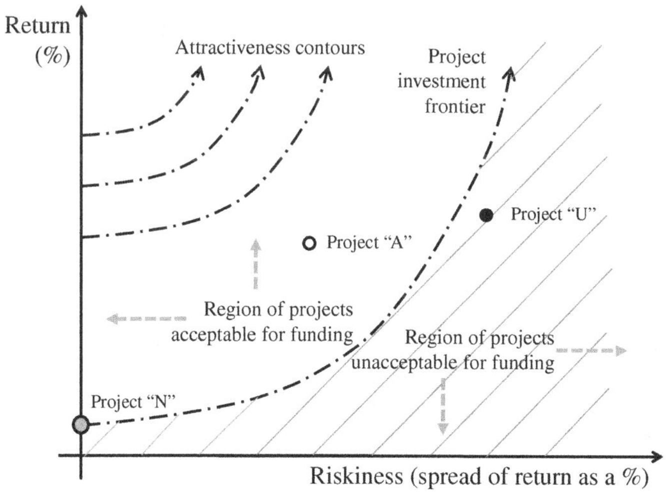

Fig. 7.4 The project attractiveness map

- The attractiveness contours (including the project investment frontier) are shown with as curving "upwards to the right". This reflects the aversion to risk we assume for a project funder.

There are, of course, other reasons that a project might be rejected for funding even if it fell into the upper left region, such as being inconsistent with the organisation's mission statement or strategic guidelines. In this discussion, we confine our attention to those projects that are acceptable against all other criteria, so that we can focus on the relationship between return and riskiness.

As well as indicating whether a proposed project is acceptable for funding, this diagram can also be used to determine (conceptually) whether a completed project was successful (see Chap. 8).

7.3.2 Calculating the Expected Return of a Project

Section 7.1.4 introduced an illustrative project in which a pharmaceutical company (LPC) undertakes an initiative to market a genomic vaccine (GVE). Based on the various analyses of this project earlier in this chapter, Table 7.5 provides a summary of the key variables required to develop the attractiveness maps discussed below.

This analysis suggests that the implementation of the risk mitigation programme *increases* the expected return from the project from 195 to 200%, while at the same time reducing its riskiness from 40 to 22%. Once again, it must be noted that a risk management plan *which just happens to reduce* a project's return may still be adopted if, at the same time, it increases the project's attractiveness, as is discussed next.

Given the calculation in Table 7.5, we can now plot the project on the attractiveness map (see Fig. 7.5), for example to allow a comparison with other project proposals competing for the same budget. With a "post RMP" riskiness of 22%, the business case for the new genomic vaccine project is represented in Fig. 7.5 by point B (22, 200). Assume that the business case contour (which, by definition, passes through point B) cuts the return axis at 150%— the threshold ownership return set by the funder for all prospective projects—as represented by point E (0, 150).

Table 7.5 A summary of the data used to analyse the return and riskiness for LPCs genomic vaccine project

Variable	Anticipated values (%)
Expected return (in the absence of any risk mitigation)	195
Riskiness (in the absence of any risk mitigation)	40
Expected return (based on expected project worth following risk mitigation)	200
Riskiness (following risk mitigation)	22

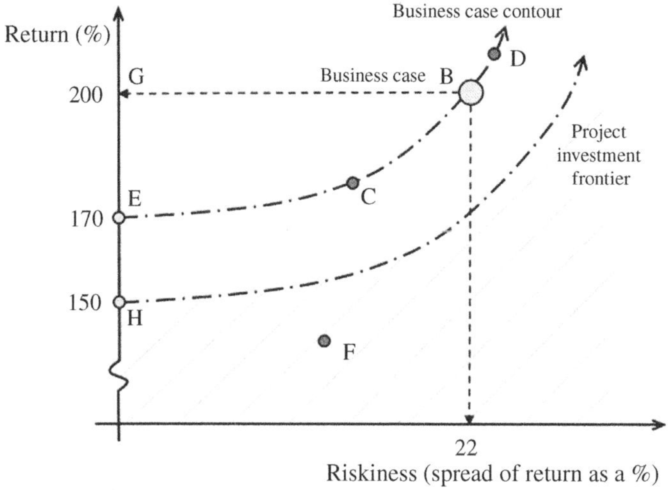

Fig. 7.5 The project attractiveness map for the genomic vaccine example

Some other important observations about Fig. 7.5 include:

- B (22, 200) shows the location of the project as defined in the business case.
- G (0, 200) indicates an expected return of 200% (if the project had been risk free).
- C and D represent projects that the funder would rank as equivalent to B in terms of attractiveness for investment.
- E (0,170) is where the business case contour cuts the Y-axis. Like C and D, E would be ranked by the funder as (in this case a *risk free*) equivalent of the project defined in the business case.
- The segment of the return axis lying below E defines all completed projects whose attractiveness *falls short* of the business case B.
- H (0, 150) indicates that the minimum return required by a funder for investment in a riskless project is 150%.
- F indicates a project that is unsuitable for funding (because it lies beneath the project investment frontier).

7.3.3 The Effect of Risk Mitigation on Project Attractiveness

Consider the following threat to a project:

- Pre-Severity = $25.00
- Pre-Likelihood = 0.4
- Pre-Expected Damage = $10.00 (=$25.00 * 0.4)

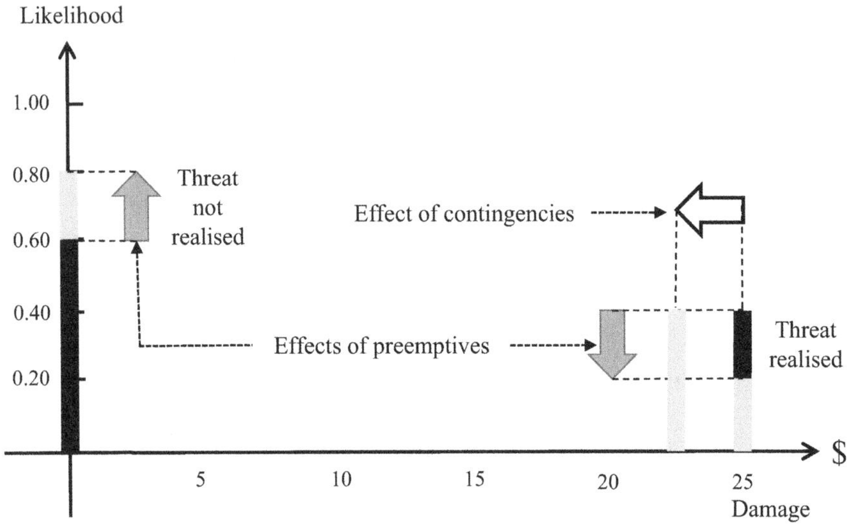

Fig. 7.6 The separate effects of preemptives and contingencies on the Bernoulli distribution of a threat

Two mitigating actions are available (one preemptive and one contingency). The preemptive causes likelihood to fall from 0.4 to 0.2, while the contingency causes severity to fall from $25.00 to $22.50:

- Post-Severity = $22.50
- Post-Likelihood = 0.2
- Post-Expected Damage = $4.50 (=$22.50 * 0.2)

Figure 7.6 shows the separate the effects of the preemptives and contingency elements of this mitigation on the Bernoulli distribution of a threat. Figure 7.7 then shows the overall effect of the proposed risk mitigation programme on the expected damage of the threat.

By definition, a preemptive causes the likelihood of a threat to fall. This is reflected in Fig. 7.6 by a reduction in the height of the $25 spike, and a corresponding increase in the height of the $0 spike. By way of contrast, a contingency causes the $25 spike

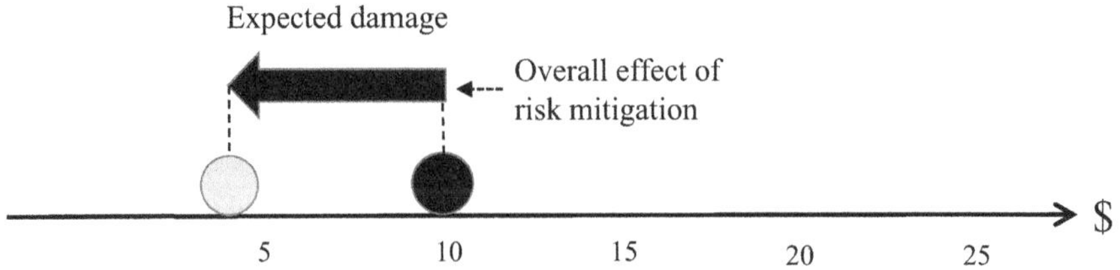

Fig. 7.7 The overall effect of risk mitigation on the expected damage of a threat

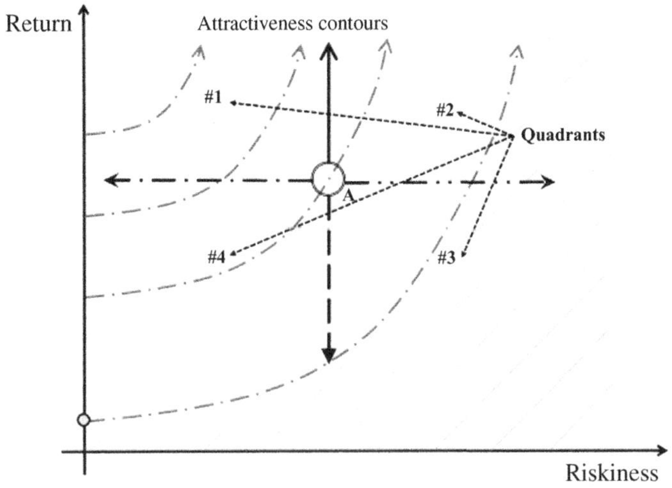

Fig. 7.8 The overall effect of risk mitigation on the expected damage of a threat

to move to the left by $2.50 (to $22.50). There is no corresponding change in the height of either spike.

While risk mitigation reduces a project's expected damage, it also incurs additional cost. As project costs rise, expected worth falls, and thus, the net impact of risk mitigation can be an increase or a decrease in project worth. Why might a decrease in project worth be accepted by a funder? If the reduction in riskiness more than compensates for the loss of project worth, then a funder may well decide to accept the proposed mitigation actions, which, in turn, reduces the worth of the project.

Taken together, the change in expected damage and reduction in riskiness have the effect of shifting a project's location on the attractiveness diagram. Consider Fig. 7.7 which shows project "A" before adoption of a proposed programme of risk mitigation. Following risk mitigation, the project will move to one of four quadrants according to the effectiveness of the risk management programme and its impacts on cost and riskiness—as suggested by Fig. 7.8—where:

1. The effectiveness of the risk management programme exceeds its cost (implying that the expected worth of the project—and consequently its return—an increases). The risk management programme causes riskiness to fall as well. All combinations of return and riskiness in this quadrant represent projects that are *more attractive for investment than "A"*. Such a risk management programme will always be adopted.
2. The effectiveness of the risk management programme exceeds its cost (implying that the expected worth of the project increases). The risk management programme causes riskiness to rise as well. As with quadrant #4, some combinations of return and riskiness in this quadrant represent projects that are more attractive

for investment than "A", but others are less attractive. The funder must trade off the increase in riskiness against the rise in expected project worth.

3. The effectiveness of the risk management programme falls short of its cost so implying that the expected worth of the project decreases). At the same time, the risk management programme causes riskiness to rise as well. All combinations or return and riskiness in this quadrant represent projects that are *less attractive for investment than "A"*. Such a risk management programme would never be adopted.
4. The effectiveness of the risk management programme falls short of its cost (so implying that the expected worth of the project decreases). At the same time, the risk management programme causes riskiness to rise as well. Some combinations of return and riskiness in this quadrant represent projects that are more attractive for investment than "A", but others are less attractive. The funder must trade off the reduction in riskiness against the fall in expected project worth.

Reflections

Box 7.4: Cost–Benefit Analysis and ITO-based Project Assessment

The economics discipline has assembled not only a formidable suite of decision-making tools, but also in a particularly robust-theoretical framework that underpins those tools. Cost–benefit analysis identifies a class of technique, which seeks to answer the question "How does the value of beneficial effects from the project, on one hand, compare with the value of the resources it consumes and any damaging effects on the other?". The answer does not take the form of a decision but is instead critical information that should be used when making a decision about a proposed initiative. A complicating factor arises from the fact that this question will have different answers depending on whose view is taken of benefits, resources and disbenefits. In general, cost–benefit analysis does not value the project from the perspective of the funder, but rather from the perspective of the entire community of stakeholders in the exercise. From this, we can conclude that while the respective roles of cost–benefit analysis and ITO-based project assessment are similar, only the latter is relevant to the project funding decision.

7.4 A Project Portfolio

A popular view is that projects, programmes and portfolios represent a three-level scale of complexity or maturity. In the discussion about programmes in Chap. 3 we have suggested that programmes are little more than collections of coordinated projects, but the concept of portfolios bears some further exploration. A portfolio is simply the collection of projects that the organisation has accepted for funding. There is no requirement that projects within a portfolio be somehow linked or related.

7.4.1 The Project Portfolio Selection Problem

When organisations have more project proposals than the available funding can support, they face a portfolio selection problem, which can be expressed as a question: "Which of a proposed list of candidate projects should be funded?" To answer this question, the funding organisation must:

1. Classify all candidate projects as suitable/unsuitable for funding (thus creating a dichotomy). This is done conceptually using the attractiveness map where projects that lie above (and to the left) of the project investment frontier are suitable for funding.
2. Rank all candidate projects from the "suitable" set in terms of their attractiveness on the attractiveness map. Again, this is done by noting where projects lie with respect to the attractiveness contours.

It is clear that the selection will be constrained by the funds to which the organisation has access. If the combined outlays required by the set of suitable projects fall short of the available funds, then all of them will be accepted, otherwise, only some of the suitable set can be funded. In that latter situation, the selection of the projects to be funded is not a trivial matter because projects are "lumpy". In some situations, it could well be better to fund two low-ranked candidates than one high-ranking alternative. This is known as the Knapsack problem—which, in general, is particularly difficult to solve. A useful heuristic that is easily applied, but which will, in general, yield suboptimal portfolios, involves the following steps:

1. Locate candidates on the attractiveness map (using their return and riskiness values).
2. Delete from consideration of all those proposals that fall below the project investment frontier.
3. Rank those that remain in attractiveness order.
4. Compare the cost of the first project in the list with the remaining budget. If the remaining budget is adequate:

    a. Add this project to the portfolio.
    b. Remove the candidate from the list (if the list is empty, stop the process).

c. Reduce the remaining budget by the cost of the project that has just been selected.
d. Repeat this step until the remaining budget is inadequate for any of the remaining candidates. In such a case remove the candidate from the list and check the next proposal in the ranking order. If the list is empty, stop the process.

When viewed in this way, project portfolio management is based firmly on the foundations of modern investment portfolio theory, which seeks to maximise the return on a set of assets (such as shares). Investment theory identifies the most attractive set of assets based on their expected return on investment and risk. In project investment management, we use the concept of "return" (as discussed in Sect. 7.1) as a surrogate for "return".

An organisation's project portfolio is the collection of initiatives that it has decided to fund. Because of continuous change in the business environment all organisations must solve the portfolio management problem: of all the projects we face at the moment (existing and prospective) which are to be funded? Solving the portfolio management problem involves the maintenance of three lists of projects: those in the current portfolio which will continue to be funded, those in the current portfolio which are to be abandoned, those prospective projects which are to be incorporated into the portfolio.

7.4.2 Strategy Implementation Through Project Portfolio Selection

Some observations about Fig. 7.9 include:

- All points on an attractiveness contour represent projects with equal attractiveness to the funder. The funder is said to be *indifferent* between projects lying on the one contour. "E", "F" and "G" are all equally attractive for investment.
- Project "A" is not only worth more than any of the other candidates, but less risky. In decision theory, "A" is, what is known as, a "dominant" option. This means that, for a risk-averse funder, it would always be at the top of the list of candidates sorted by attractiveness *regardless of the shapes and locations of the attractiveness contours*. If, however, the project investment frontier happened to lie above "A", then none of the projects displayed would be suitable for investment.
- "K", which has the lowest attractiveness for funding, would, for a risk-averse funder, always be at the bottom of the list of candidates.
- "B" has a lower return than C, but more attractive because it lies on a higher contour (If our contours were "flatter"—so that one passed between B and C, then the reverse would be true).
- Despite its high return, "I" is unacceptable for funding (because its return is inadequate compensations for its high riskiness). Similarly, "J" and "K" would not be considered. If the project investment frontier moved down and to the right, it is conceivable that all three could eventually become candidates for investment.

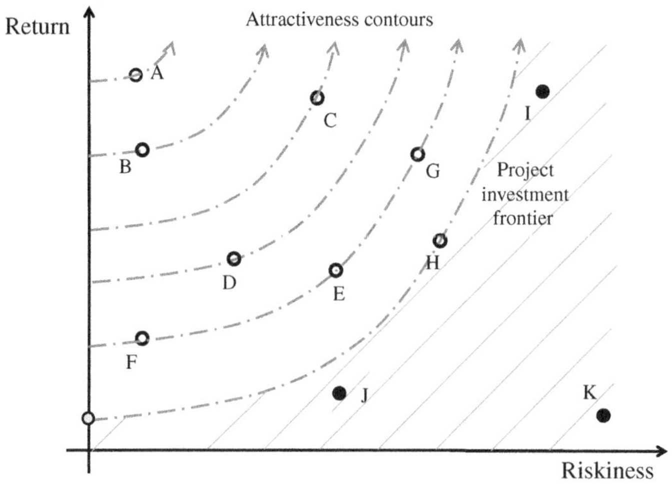

Fig. 7.9 Candidate projects for investment displayed on the attractiveness map

Summary

Not all projects are suitable for funding. Moreover, when resources are limited, only the "most suitable" project proposals should be funded. We have defined "suitability" in terms of attractiveness—which is then used to rank prospective projects. Attractiveness allows a funder not only to rank proposed projects, but to classify them as worthy/unworthy of investment. Attractiveness involves a trade-off between expected return and riskiness. Return is a function of the project's benefits, disbenefits and costs while riskiness is a statistical measure of the "uncertainty" surrounding the achievement of a specified level of return. Return and riskiness can be usefully displayed on an attractiveness map with return on the vertical axis and riskiness on the horizontal axis. All this enables an organisation to compare projects and assemble a portfolio—the collection of projects that have been adopted for funding. Armed with the tools that underpin attractiveness, in the next chapter, we assemble a comprehensive analytical model to gauge the success of a completed project.

Chapter 8
Project Success

Abstract Accepted wisdom holds that a project is successful if its outputs are delivered fit-for-purpose, on time and within budget. If, however, as is implied by the Input-Transform-Outcome (ITO) model, beneficial outcomes reflect a project's purpose, then the conventional view is inappropriate (because it judges success without reference to the realisation of target outcomes and hence the eventual generation of benefits). In this chapter, we introduce an assessment framework for projects and explore some key issues surrounding its application. According to this framework, "project success" involves separate judgements across *three distinct layers*—not just one. Two of these relate to the performance of two key players—the project manager and the project owner, respectively. The third relates to the performance of the investment represented by the project from the funder's point of view. This approach to judging success employs the concepts of project "worth" and "return" introduced in Chap. 7. We also review the results of some research into the way project success is treated in practice and go on to consider some important determinants of success—in particular the concept of Critical Success Factors (CSF). The discussion examines some limitations of that concept and investigates an alternative approach, entitled Critical Success Processes (CSP).

8.1 The Three-Layered Model of Project Success

This section lays the foundations for a comprehensive treatment of project success. The same foundations will also be used in Chap. 9 to explore the processes that surround the decision to approve a proposed project. Although these two topics concern "opposite ends" of a project, they share the same conceptual model.

Success is the state of having achieved an objective, subject to a set of accepted constraints. To judge the success of any process, it is necessary to evaluate its performance by measuring one or more selected variables and then comparing each observation with a preset threshold value. In other words, success is judged after the performance has been evaluated.

The literature reveals a wide range of views about project success. Baker et al. (1988) stated that there is no "absolute" success but only "perceived" success, imply-

O. Zwikael and J. R. Smyrk, *Project Management*,
https://doi.org/10.1007/978-3-030-03174-9_8

ing that project success means different things to different stakeholders at different times (Shenhar and Holzmann 2017). A project may have many stakeholders whose interests differ widely (and, in some cases whose interests fluctuate wildly over the course of events). So, whose views about a project really matter? To answer the question, we need to distinguish between two related questions (each of which has two parts, related to the front and back ends of the project, respectively):

1. "Why is it necessary to make judgements about a business case and (later) about the completed project?"
2. "Whose views about a project *should* be used to accept or reject a project and guide judgements about its eventual success?"

As far as the first question is concerned, judgements have to be made about a business case (ex ante) so that a decision can be taken on funding the proposed initiative. Judgements about a completed project, on the other hand, are required (ex post) to decide if the accountabilities held by the project owner and the project manager have been discharged and to determine if the project represents a satisfactory investment.

The second of the questions posed above can be addressed using the ITO framework introduced in Chap. 2. Clearly, the funder's view alone will determine acceptance/rejection of a project proposal (because he/she controls the related budget). In a similar line of reasoning, it is the funder's view alone that determines the eventual acceptability of the resulting investment. A further issue bears discussion at this point: "Should the views of other stakeholders be considered when making decisions about projects?". Clearly, only the funder can provide a satisfactory answer to the question and so we can conclude that all views about a project are subordinate to those of the funder (In fact, the entire stakeholder management process covered in Chap. 5 is based on this conclusion).

After project approval, the funder will have made a project owner accountable for achieving a level of return derived from the data provided in the business case. This implies that the funder must eventually make a judgement about the project owner to decide if he/she has successfully discharged that accountability. Likewise, before planning gets under way, the project owner will have made a project manager accountable for eventually delivering its committed outputs within the constraints imposed by the project plan. This implies that the project owner must, in turn, eventually decide if the project manager has successfully discharged that accountability.

From this discussion, we conclude in Table 8.1 that *three* judgements should be made about the success of a completed project rather than just one (Zwikael and Smyrk 2012). Other established approaches to the determination of project success appear to be special cases of this three-layered model of project success. Discussion of tests of project success requires an introduction to the underlying generic structure of a test—which follows.

As is discussed below, the tests of ownership and investment success both use the project attractiveness map (introduced in Chap. 7).

Table 8.1 Key aspects of the three-layered model of project success

Level of test	Project management success	Project ownership success	Project investment success
Who judges?	Project owner	Project funder	Project funder
Who is being evaluated?	Project manager	Project owner	The investment
What is judged?	Achievement of the project plan	Realisation of the business case	The effective "return" on the investment in the project
Relevant criteria	Time Cost Scope/quality Detrimental outcomes	Achievement of the approved business case	Acceptability of the realised business case

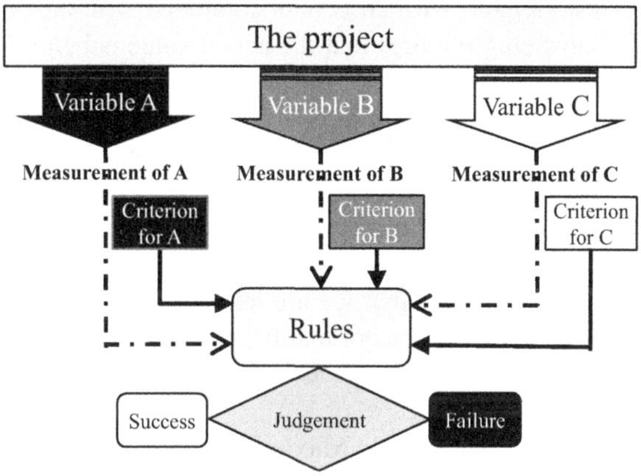

Fig. 8.1 The general structure of success tests in projects

8.2 Judging Success

8.2.1 The Nature of a Success Test

In general, a test of success is made up of a specific set of performance variables, a measurement on each of those variables, a set of criteria (taking the form of a reference value for each variable) and a structure of rules showing how to use the resulting measures to make a judgement about success. Figure 8.1 illustrates the application of a test to make judgements about a project.

Execution of a test involves the following steps:

1. Take a measurement of each test variable (e.g. the project was completed in 15 months).
2. Compare the measurement and the associated criterion (e.g. the actual project duration of 15 months is longer than the expected duration of 12 months).

3. Use the structure of rules to map the set of resulting differences onto a binary judgement: success/failure. Amongst other things, these rules will take into account the fact that the project exceeded its target duration by three months.

From the literature

Box 8.1: Satisficing as the foundation of scientific test

In Box 7.3 we referred to the work of Simon (1976) in which he introduced the concept of satisficing. Based on this concept, we assume that a funder seeks to satisfice the investment performance of a project rather than optimise it.

We assume satisficing behaviour in project funders for another reason as well. Tests of project success demand that a threshold value be set for each test variable—which is then compared with the actual value of that variable. Satisficing requires that the actual value is better (above or below depending on the nature of the variable) than the threshold. In fact, without this threshold, no judgement can be made about success. We usually know little about the response of a project's target outcomes to changes in the variables which determine their achievement. If we know nothing about a function we cannot identify its maxima or minima, therefore there is no test for optimality (such as the first derivative reaching zero).

So, the issue that we are left with is this: how are we to judge success if we cannot test for optimality? Consider these two (contrasting) situations in which a sales manager is setting (with the CEO) target outcomes for a new product initiative.

- Target outcome A: Maximised sales (with a target of "maximum").
- Target outcome B: Increased sales (with a target of 15%).

Clearly, we could not determine if the project had been successful using target outcome "A", whereas with "B" we could.

8.2.2 *Absolute Versus Trade-Off Tests of Success*

Two types of rule structure are of particular interest: absolute and trade-off (Meredith and Mantel 2015). In an absolute rule structure, each difference between a variable and its associated criterion is mapped independently onto the judgement so that the failure of *any one* variable to meet its criterion results in an overall judgement of project failure. Equivalently, if all variables meet their criteria, then the project is judged as a success. By way of contrast, a trade-off test maps all the differences collectively onto the judgement in such a way that underperformance in one variable may be traded off against over-performance in others.

Shortcomings in the absolute test become obvious by examining a number of project performance scenarios. Consider a one-year/$1 M project in which only two performance variables will be applied: duration and cost (Such a test would be of dubious practical value but suffices for the purposes of this discussion). Under scenario A, project execution is completed in ten months, but the budget is exceeded by $1000. Under scenario B, project execution is completed in 13 months, but project costs come in at only $750,000. Under the absolute test both scenarios would result in judgments of failure, and yet one could readily postulate situations where the project funder would be happy to trade-off between time and cost and regard the project as having been successful in each case. Clearly, the absolute rule structure is a special case of the trade-off structure. In practice, trade-off among success measures is commonly used by decision makers and so, without any loss of generality can be employed in all project success frameworks.

8.2.3 Assessment

Assessment can take either of two forms—distinguished according to when they are executed (Campbell and Brown 2003):

- **Appraisal** (also known as "ex ante assessment") is carried out before a project starts, to support decisions related to acceptance/rejection of the business case and (later) the project plan. Appraisal is the focus of Chap. 9.
- **Evaluation** (also known as "ex post assessment") is carried out "at the end of" a project (described in what follows), in support of judgements about success/failure. The remaining discussion in this section is related exclusively to evaluation.

Evaluations of success are made for each of three subjects: the project manager (project management success), the project owner (project ownership success) and the investment represented by the project (project investment success). Each of the three tests uses its own list of parameters.

The trigger for a test of project management success *is the end of project execution* (when the last of the project's outputs has been delivered). The earliest point at which overall project management success can be judged is on delivery of all outputs. Any judgement about project management success after that will be based on the same data as was available then—except for any new detrimental outcomes that come to light later. A detrimental outcome is an undesirable outcome attributable to a decision or action of the project manager with either of two characteristics:

- It was anticipated, but subsequently experienced at higher levels.
- It was not anticipated, and subsequently experienced at unacceptable levels.

While unlikely, it is conceivable that some detrimental outcomes emerge during outcomes realisation (after execution). Because of this, a judgement about project management success made at the end of execution is *conditional* on nothing untoward being discovered later. In the case of a contracted project manager, such an eventuality would have to be covered in the commercial framework governing the appointment.

The trigger for a test of project ownership success is the end of outcomes realisation—defined as the point at which the flow of project returns has been secured. Project returns are "secured" in any of two situations:

- The flow of project returns achieves the threshold set by the funder in light of the approved business case. Furthermore, over an acceptable timeframe, that flow is forecast not to fall below the threshold.
- The flow of project returns does not achieve the threshold set by the funder in light of the approved business case. Furthermore, over an acceptable timeframe, that flow is forecast not to reach the threshold.

Project ownership success is judged when the flow of project returns has been secured. Any judgement about project ownership success after that will be based on exactly the same data as was available then (nothing that impacts project ownership success can change after outcomes realisation) and so judgements about project ownership success are also insensitive to their timing.

The situation with project investment success is, however, subtly different. While a judgement about project investment success is also based on the actual values of project returns, the forecasts will vary according to the point at which they are made and so judgements about project investment success are sensitive to their timing.

8.2.4 The Project Investment Paradox

Research points to significant overall rates of failure in projects (Zwikael and Smyrk 2012). At the same time, investment in projects continues to grow (Shenhar and Dvir 2007). This suggests a *project investment paradox* (Ryder-Smith 1998)—of which the *IT productivity paradox* can be viewed as a special case. The question now arises, if most projects fail, why do organisations continue to invest in them? There are a number of possible explanations:

1. The empirical results are misleading, and rates of failure are not as high as the research suggests.
2. Failure rates are high—but project funders are either unaware of this (or ignore it) when making funding decisions.
3. Failure rates are high, but in each funder's portfolio of projects the "winners" more than compensate for the "losers".

In what follows below, we conclude that the accepted conventional test used to judge projects says little (if anything) about the success of the *investment* represented by a project. This means that a project that does not achieve its business case goals, may still be judged as "successful". We also conclude that misuse of this test introduces a systematic upward bias in project failure rates. If funders suspect this and, at the same time, apply more appropriate (albeit intuitive) tests of investment success, then the first conjecture would serve as a reasonable working hypothesis.

8.2.5 Regression Testing

The question now arises "What device can be employed when making the three judgements about success?" Regret theory (Bell 1982) posits that people care not only about what they get but also about what might have been obtained had they chosen differently. Pinto and Slevin (1990) used this theory in their project success measurement: Knowing what you know now about the status of the project, would you have funded it?

Based on this theory, we propose a regression test which uses certain performance evaluation parameters found in the baseline documentation. However, the original values of the parameters (stating what was intended to happen) are replaced with their actual values representing what actually happened subsequently on the project. We distinguish between these two versions of the parameters by labelling them "approved" and "achieved". If, for example, the "approved" timeframe and cost were 12 months and $1 M, respectively, but the exercise finished up taking 15 months and $1.2 M the higher figures are now treated as the "achieved" values of the two parameters.

The regression test accepts trade-offs amongst the project's performance evaluation parameters. This principle implies that even though some of the evaluation criteria are not met, certain combinations of actual values can, in some cases, be treated as equivalent to the same combination of their achieved values. If in the previous example, the achieved business case showed actual values for timeframe and cost at 10 months and $1.1 M, respectively, the funder could well decide that they are equivalent to those appearing in the approved baseline documents (where these parameters had been set to 12 months and $1 M, respectively).

The proposed regression test is based on the understanding that success is judged by comparing *original intentions* with *subsequent achievements*. Whereas completing such a project in 15 months may be an "excellent" achievement when compared with other similar benchmarked projects in the same industry, it may well be considered poor when compared with a target of 12 months.

More formally, the regression test involves the following steps:

1. In the approved baseline document, remove the values that were set, anticipated or estimated for certain parameters including target outcomes, undesirable outcomes, costs, timings, risks and issues.
2. Replace the original values with those that were actually experienced or realised at the end of the exercise. This gives rise to the "achieved" version of the relevant baseline document.

3. Determine the equivalence of the approved and achieved baseline documents by comparing just those parameters that are relevant to the test under consideration (project management, project ownership or investment)—as discussed in the following sections.

8.2.6 Which Version of the Performance Evaluation Parameters Should Be Used in Regression Testing?

There is a practical issue to be considered when applying the regression test to project management and project ownership. When setting the threshold for an acceptable level of project return, what version of the business case is to be used? While at first this seems obvious (use the values set in the original), versions, the situation becomes less clear when one considers that all of this detail will be the subject of continuous revision from the point when it was originally accepted through to the project end. A pragmatic approach to this issue is to base the regression test on the last approved version (with progressive minor changes being ignored).

8.3 Project Management Success

In this section, we examine the conventional test of project management success, offer a critique of the underlying approach and discuss the alternative framework introduced above.

8.3.1 The "Iron Triangle"

The established test for project management success is often framed in terms of the "iron triangle" (Jha and Iyer 2007). This test evaluates the extent to which the committed outputs from the project are delivered subject to three criteria:

1. **Timeframe**. By an agreed date.
2. **Cost**. Within an agreed budget.
3. **Scope/quality**. Delivery of all committed in accordance with the level of quality specified for each.

In the most common form of the test, the three criteria are all applied absolutely and independently. In other words, failure of any one criterion implies failure of project management. As discussed in Sect. 8.2.2 however, another variant allows trade-offs amongst the three criteria as explained in Table 8.2.

The iron triangle may well have its genesis in projects that are defined by *fixed price* contracts. A bid for a job commits the contractor to criteria for output quality, timeframe and a *price* (the amount to be invoiced). "Price" here refers to the monetary value of the outputs covered by the contract. The price is in turn based on an estimate of the *cost* incurred in meeting the quality and timeframe criteria. "Cost" here refers to the monetary value of the resources consumed by the work undertaken by the contractor to produce those outputs. In general, from the contractor's perspective, a project will have been successful if the criteria for scope/quality, timeframe and cost have been met.

Despite its wide acceptance, three criticisms can be levelled at the way the iron triangle is applied in practice. First, it is not entirely clear that the issue raised in Sect. 8.2.6 is being satisfactorily addressed. In other words, when the results of its application are reported rarely, if ever, is it stated which version of the project plan has been used as a baseline for this comparison. A second issue arises in those cases where it is not stated whether the absolute or trade-off version of the test is being applied. Declarations of success/failure made without confirming that are difficult to validate. Finally, it would appear that the iron triangle fails as a test of project management success because it ignores *detrimental* outcomes. The "digital industrial pressure controller" illustration (Box 8.2) exposes such a weakness. This issue is discussed further in Sect. 8.3.2 below.

Others have also recognised the problems that arise when attempts are made to apply the iron triangle to tests of project success, suggesting that it is partial, unsatisfactory and misleading because it ignores important "soft outcomes", such as the satisfaction of the client or employee development (Scott-Young and Samson 2008). We mention in passing that classifying outcomes as "soft" carries with it a value judgement that they are, in some sense, of less significance as success criteria than, say, time and cost. The ITO model implies that success criteria cannot be

Table 8.2 The two variants of the iron triangle

	Variant	
Assessment test component (Fig. 8.1)	Absolute test	Trade-off test
Criteria	Thresholds of all relevant parameters set in the project plan	All relevant parameters set in the project plan
Rules	If the threshold for each variable is achieved, then project management is successful If threshold for any variable is not achieved, then project management is unsuccessful	Given delivery of all committed outputs to the agreed quality, if the relative levels of performance against time and cost are acceptable then project management is successful Given delivered scope/quality, if the relative levels of performance against time and cost are unacceptable then project management is unsuccessful

meaningfully ranked in this way and that they are of equal value when applied to tests of success.

Illustrative example

Box 8.2: The Digital Industrial Pressure Controller Project—A Failure of the "Iron Triangle" as a Test of Project Management Success
Consider a project undertaken by a specialist instrument manufacturer (identified here as "SIM Inc."), in which a new digital industrial pressure controller has been developed under contract for a new client who would otherwise have bought a similar number of a current analogue model.

The project manager delivered a prototype of the new instrument in accordance with its specification, on time and within budget, but achieved this by putting the team under such pressure that a number of key technical staff resigned towards the end of the project. How is SIM Inc. to judge the management of the project (and, by implication, the project manager)? Based on the iron triangle a judgement would have to be made that the project was successful (because all three criteria have been met), but if the loss of staff was considered undesirable, unexpected, avoidable and unacceptable, then *the management of the project* might well be declared unsuccessful. This conclusion implies that the iron triangle fails as a test of project management success, because it ignores detrimental outcomes (those undesirable outcomes that can be attributed to the management of the project).

8.3.2 The "Steel Tetrahedron"

We have raised three criticisms of the conventional test of project management success, one of which concerns its failure to recognise undesirable outcomes. To deal with this, we propose a variant of the regression test in which the three accepted performance evaluation parameters are augmented with a fourth:

1. **Timeframe**. By an agreed date.
2. **Cost**. Within an agreed budget.
3. **Scope/quality**. In accordance with an agreed specification that identifies all of the deliverables and sets the required level of quality for each (as a list of specified attributes).
4. **Detrimental outcomes**. Unanticipated, undesirable outcomes that are attributable to decision or actions by project manager (These include undesir-

Table 8.3 A hypothetical scenario at the end of the digital industrial pressure controller project (Box 8.2)

	Relevant values of project parameter	
Project management success parameter	Targets established in project plan (as thresholds)	Actual values achieved at the end of execution
Scope/quality	Set in the business case	All outputs delivered fit-for-purpose
Cost	\$2.5 M	\$3 M
Timeframe	30 months	24 months
Undesirable outcomes	Loss of two key staff	Loss of one key staff member

able outcomes that were completely unanticipated anticipated but experienced at higher actual levels than initially expected).

We can conclude, therefore that *four* success measures are necessary and sufficient in a test of project management success, and so we suggest (slightly tongue in cheek) replacing the ubiquitous triangle with a tetrahedron, as suggested in Fig. 8.2.

It should be noted that the use of geometric shapes to relate the various criteria has no analytical significance whatsoever—they are purely pictorial devices. Table 8.3 illustrates the use of these measures for a hypothetical situation that may arise at the end of the digital industrial pressure controller project in which (for simplicity of discussion) we will assume that all outputs were delivered fit-for-purpose.

When using the regression test to evaluate the performance of the project manager, the project owner would note that the original committed outputs had been delivered to their agreed specifications and so is now concerned only with the other three vertices of the steel tetrahedron—those involving cost, time and detrimental outcomes. To do that, the project owner must make a judgment about whether or not the set of actual results in Table 8.3 is at least "equivalent" to the set of target results specified in relevant parts of the business case and the project plan.

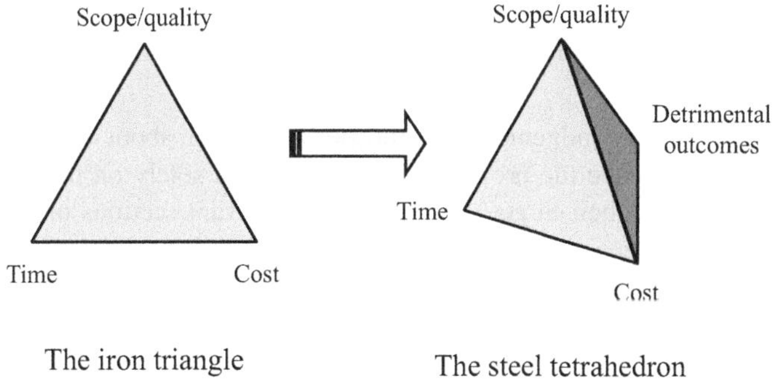

Fig. 8.2 Gauging project management success: from the "iron triangle" to the "steel tetrahedron"

This discussion implies that only those elements of the four measures that can be attributed to the decisions and actions of the project manager can be taken into account when evaluating project management success, as illustrated by the following three examples:

- **Time overruns**. Extraordinary demands for heavy lift marine cranes in China slow the rate of delivery of precast sections of a new concrete box girder bridge in Malaysia, causing significant delays in the opening of a new motorway. If such an event was judged unforeseeable, then the project manager could not be held responsible for the failure to meet the schedule. If, on the other hand, there had been considerable discussion about possible accelerated investment in Chinese port infrastructure, then a view might well be taken that such a threat should have been recorded in the bridge project risk register and that the project manager failed by not recognising the risk (Such a view could well be taken of any other threats that the project manager should have identified but missed).
- **Unanticipated undesirable outcomes**. Only those that can reasonably be treated as "caused" by the project manager are relevant to judgements of project management success. If, in the case of the digital pressure controller (Box 8.2) it is discovered that, in addition to the anticipated cannibalisation of the old analogue product, sales of another multifunction product were significantly reduced as well, then one would be hard pressed to "blame" the project manager. Cannibalisation of existing products (whether expected or not) is certainly an undesirable outcome, but one that has nothing to do with the management of the project.
- **Anticipated undesirable outcomes that have been identified in the business case**. Two situations need to be considered here:
  - Where any undesirable outcomes that have already been identified and anticipated are experienced at lower levels than anticipated in the business case. Here project management success is *unaffected* by such an eventuality.
  - Where any undesirable outcomes that have already been identified and anticipated are experienced at *higher* levels than anticipated in the business case. Here project management success *is* affected by such an eventuality.

8.3.3 Judging Project Management Success

The regression test for judgements by the project owner about the management of the project (and hence the project manager) is based solely on the four criteria mentioned above and their target values as set in relevant sections of the baseline documents. Any changes in all other parameters (such as target outcomes) lie outside the responsibility of the project manager and hence can play no part in judgements about the discharge of his/her accountability.

A comment about the risk register is necessary. How is the project owner to assign accountability for risk to the project manager? Two positions are worthy of comment. The first is that, beyond meeting that part of the project budget allocated to the risk

Table 8.4 Project management performance levels across industries

Industry type	Schedule overrun	Cost overrun	Quality of outputs
Scale	%	%	1–10
Engineering	13.6	10.5	8.3
Software	14.5	10.5	7.4
Production/maintenance	18.4	11.9	7.5
Construction	12.8	7.7	7.9
Telecommunications	14.9	10.6	8.1
Services	10.5	8.8	7.9
Government	26.0	14.9	7.6
NGO	7.7	4.7	7.3

management programme, the project manager has no accountability. In other words, the realisation of threats has no impact on the way the project owner judges the performance of the project manager. The second is that the project manager is held accountable for holding the overall level of damage from realised *threats to no more than the expected damage revealed by the risk register*. We propose that the second of these approaches be adopted because the project manager then has an incentive to minimise the damage from realised risks. We can, therefore, conclude that the project manager is accountable for achieving the project plan in general and for limiting the damage from realised risks to the level of the expected damage revealed by the risk register.

8.3.4 Project Management Success Rates in Practice

The purpose of this section is to examine the extent to which organisations meet project management success criteria. Projects that did not meet their targets are well known. For example, projects that have significantly exceeded their approved budget include the Suez Canal, Egypt (1900%), Scottish Parliament Building, Scotland (1600%), Sydney Opera House, Australia (1400%), Concord Supersonic Aeroplane, UK, France (1100%), Troy and Greenfield Railroad, USA (900%) and Montreal Summer Olympics, Canada (720%) (Flyvbjerg 2017). On the other hand, some prominent projects have been completed on time, for example, New York's Empire State Building, the Paris' Eiffel Tower and the Guggenheim Bilbao Museum in Spain (Flyvbjerg 2005).

A study conducted by the authors analyses the three success criteria for 2002 projects across eight industries in different countries. Table 8.4 compares project management performance in terms of meeting targets set for: schedule overrun, cost overrun and quality of outputs. The table shows the mean values in each case.

If evaluated using the absolute variant of the conventional test for project management success (involving separate criteria for scope/quality, time and cost), then Table 8.4 suggests "poor" results for most projects. Given our earlier criticisms of the "iron triangle" test, it is necessary to qualify this conclusion. First, it does not account for undesirable outcomes. To the extent that they have been realised in individual projects, then the "true" rates of failure would be *at least as great* as indicated here. Second, this study is not based on a regression test for each of the projects included in the sample (it uses absolute criteria where trade-offs have not been accommodated). A regression test of that kind would indicate "true" rates of failure *no greater than* indicated here. On the assumption that undesirable outcomes and trade-offs are no more common in one sector than another, we can conclude that the *relative performances* of the sectors shown in this table are valid.

Table 8.4 illustrates that construction and engineering organisations have comparatively low schedule and cost overruns. The relatively good performance of these industries across all project management success dimensions is worthy of note. A number of factors may explain this result. Construction and engineering projects are subject to particularly thorough planning, analysis and review. This preparation is often concerned with an examination of numerous options before a preferred approach is finalised. While scope can and does change in engineering and construction, anecdotal evidence suggests that, because the scope is normally well defined at the outset, it rarely has to change in response to scoping errors. When it does occur, scope change is treated very seriously and managed in a thorough, systematic and formal way.

In particular, a number of explanations can be advanced for this high level of capability across the engineering, construction industries:

- These are mature disciplines which can trace their lineage back for hundreds (if not thousands) of years.
- Project management is recognised as a specific well-defined sub-profession.
- Project managers usually go through a lengthy apprenticeship working as a member of project teams.
- Over the years, a culture has emerged in which standards, methodologies, practices and procedures are encouraged, valued and accepted.

By way of contrast, the software sector, for example, achieves relatively low scores. This field is characterised by significant technological uncertainty, rapid change, and high staff turnover. Projects are subject to frequent changes of scope (often to deal with problems arising from poor initial scoping). Another factor may explain much of the relatively poor performance of this sector when compared with engineering and construction. Software development appears to be characterised by labour-intensive "hand-crafting" processes, each of which tends to involve a high level of novelty. In engineering and construction work, many outputs emerge from assembly-type processes in which components are bought off-the-shelf and "put together" (possibly after modification). Not only are the costs of the components known, but also the assembly processes are relatively standardised. If this conjecture is correct, then the ratio of labour/output-value for projects undertaken by the software industry will be significantly higher than for the engineering/construction sector. In that case, estimation of outlays for purchased components involves only counting and pricing. Estimation of labour involves judgements about work. Because the reliability of labour estimates is inversely related to the novelty of the work being estimated, we can conclude that, in general, project estimates for the software sector will be less reliable than those for engineering/construction. Interesting evidence for the low rates of project management success in this sector can be found in reports conducted by Standish Group, known also as the "Chaos report". For instance, organisations completed only 39% of software projects on time and budget, which led to estimated annual "losses" for the United States and European Union markets of around US$100 billion each (Symons 2010).

It is difficult to draw any useful inferences from these results for levels of project management success in the IT/IS sector (for which there is a dearth of empirical evidence). Two lines of a priori reasoning lead, unfortunately, to opposite conclusions. On the one hand, one might argue that while formal business case regression testing may reveal higher levels of success with projects than with their management, the monotonous procession year-by-year of studies reporting extraordinary rates of failure offer little hope that such a situation actually prevails. On the other hand, we see a long history of ongoing investment by organisations in new IT/IS-related projects. Presumably, senior management is making considered judgements to proceed with these ventures and is, therefore, being encouraged by acceptable "true" underlying rates of success.

The general perception of projects undertaken in the IT/IS sector is that their success rates are low. This view is supported by similar results from other studies. For example, an investigation of business projects with significant IT outputs undertaken by mortgage firm (Tichy and Bascom 2008) revealed that 11% were cancelled, 78% challenged (completed but unsuccessful) and only 11% successful. Poor management is considered as a major reason for these disappointing results, as well as poor leadership, stakeholder management, estimation methods, risk management and executive management support.

In summary, while numbers may vary by industries and type, the evidence suggests that many projects do not achieve the criteria for project management success.

8.4 Project Ownership Success

Just as the project manager is evaluated by the project owner, the project owner is evaluated (by the funder). Whereas judgements about the performance of the project manager are based on only the values of scope, timeframe, cost and detrimental outcomes set out in the project plan, the performance of the project owner is based on the achievement of the project's level of attractiveness as defined in the last approved business case.

8.4.1 *Judging Project Ownership Success*

At the end of initiation, when a project is accepted for funding, the funder is essentially saying “I am making an investment in a which appears on the attractiveness map in a position defined by the business case. I will use this to set a threshold—for which I expect to see an equivalent return realised when the project is finished”. Such an expectation is consistent with the concept of satisficing discussed in Box 7.3.

When accepting the business case, the funder commissions the project owner to act as his/her agent—making that person accountable for realising certain parameters appearing in the business case which enable it to be illustrated on the attractiveness map. These parameters (drawn from Sect. 7.1) are summarised in Table 8.5.

Effectively the funder is saying to the project owner “I am funding a project to achieve a certain threshold value of project return. This value is based on data you have provided me in the business case”. Eventually, a judgement will have to be made about how successfully the project owner has discharged that accountability.

A regression test applied to the calculation of project attractiveness provides the foundation for judging the performance of the project owner. It should be noted that, because fortuitous outcomes are not identified in a business case, they play no role in the test of ownership success.

8.4.2 *Using the Project Attractiveness Map in the Test of Project Ownership Success*

In essence, the regression test of ownership success turns on the question of whether the achieved attractiveness of the project is *at least equivalent* to the threshold set by the funder when the business case was approved. The project attractiveness map can be used to confirm this equivalence. As discussed in Chap. 7, while proposed

Table 8.5 The determinants of attractiveness for the genomic vaccine initiative

Parameter	Value ($000)	Source
Benefits	$2,015,000	Derived
Disbenefits	$290,000	Derived
Project costs	$540,000	Estimated
Expected damage (following risk mitigation)	$63,423	Risk register
Expected project worth (following risk mitigation)	$1,108,576	Derived
Expected return (based on expected project worth following risk mitigation)	200%	Derived
Threshold of return by which funder will judge the eventual performance of the owner	170%	
Riskiness (following risk mitigation)	22%	

projects can lie anywhere in the attractiveness map, completed projects can appear only on (or "close to") the Return axis (because most risk has evaporated during execution and outcomes realisation).

Consider the genomic vaccine example (Box 7.2). The investment is represented in Fig. 8.3 by point B (22, 200), that is, a riskiness level of 22% and return of 200%. Assume that:

- The business case contour cuts the return axis at 170%
- On completion, the project generated an actual return of 165% (shown as point A).

When judging the project owner, the funder would have to accept that the achieved return in fact fell short of the approved return because the project finished up on the Y-axis below point E. Accordingly, the funder would judge the project owner as unsuccessful (and, by implication, the project as an ownership failure).

8.4.3 *Project Ownership Success Rates in Practice*

Studies into rates of project success have found that most projects fail (Zwikael et al. 2018). For example, the Channel Tunnel has a rate of return on investment of negative 14% (Flyvbjerg 2017).

In a global research study, we analysed project success in more than 700 projects across seven industries. Because target outcomes were not identified in most of these projects, a surrogate was used—the "funder's satisfaction with the project's results". Funder's satisfaction was found to be the most important project success criterion-rated at almost twice the importance of the iron triangle (Dvir and Lechler 2004). Figure 8.4 shows the mean values of funder's satisfaction from projects across seven industries on a scale of 1 (low) to 10 (high level of satisfaction).

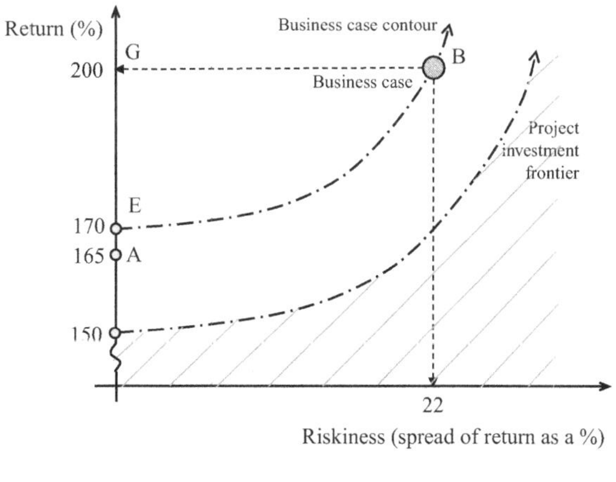

Fig. 8.3 The project attractiveness map for the genomic vaccine example

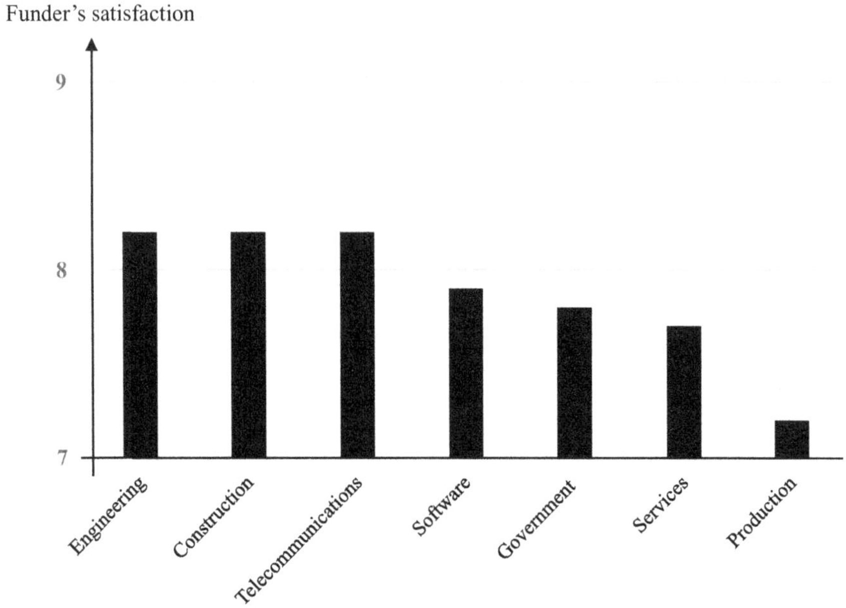

Fig. 8.4 Project ownership success in different industries

This graph suggests significant differences in project ownership success rates across industries.

8.5 Project Investment Success

Because projects involve the allocation of scarce economic resources that the funder could have employed elsewhere, it is appropriate to determine if that investment has been, in some sense, "satisfactory".

8.5.1 Judging Project Investment Success

At the conclusion of a project, the funder will pose an obvious question: "Was this a worthwhile investment?". To answer this, we need to discuss the way in which the acceptability of the business case is decided by the funder. Because acceptance of a project's business case represents a go/no-go investment decision, the question now arises: "When is a business case unacceptable?". A business case is unacceptable when the attractiveness of the project it defines falls below some predetermined threshold.

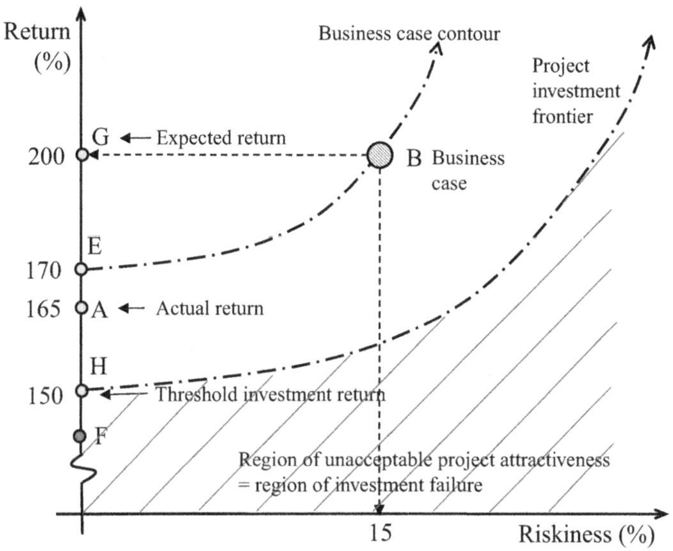

Fig. 8.5 The project attractiveness map for project investment success

Again, the regression test serves as the basis for judging the actual performance of the investment represented by the project.

Consider the attractiveness map for the genomic vaccine project (Fig. 8.3). Furthermore, assume that:

- The funder has set a global threshold value for the return of a riskless project at 150%—implying that any business case for which the equivalent expected return of less than that will not be funded.
- At its conclusion, the actual return was 165%.

Because the project's actual worth is above the minimum investment threshold, then the funder would judge the exercise as an investment success. Note that although investment failure automatically implies ownership failure, investment success does not automatically imply ownership success.

8.5.2 Using the Project Attractiveness Map in the Test of Project Investment Success

The regression test of investment success compares the attractiveness of the project defined by the achieved business case with the threshold set as an investment criterion. Once again, the project attractiveness map is used for this purpose, as displayed in Fig. 8.5 where:

- A lower contour is displayed as the "Project investment frontier"—which cuts the return axis at "H". Along this curve are all those combinations of worth and riskiness which the funder has set as the threshold for investment. Any business case which sits on or above this contour is acceptable for funding, while those below are not. The segment of the return axis lying above (and including) "H" defines all completed projects whose attractiveness exceeds the threshold set for an acceptable investment and which therefore also represents project investment success.
- The business case contour for project "B" is also shown.
- "E" represents a completed project with an attractiveness that is equivalent to the project represented by the business case.
- Note that although a project with an actual return of, say 165% "A" represents an investment success, it is, at the same time an ownership failure.

The discussion in Box 8.3 presents possible conclusions that might be drawn from a project investment success assessment.

Practicalities

Box 8.3: The Regression Test of Investment Success—The *Correctness* of a Decision versus its *Appropriateness*

The regression test used to gauge investment success is based on an assessment of the end result of the original funding decision. While we use the regression test to judge the performance of the investment (and hence the "appropriateness" of the decision to fund the project), we cannot use it to judge "correctness" of the original decision. We have here two independent concepts surrounding a funding decision: its "correctness" and its "appropriateness".

A correct decision is one in which the decision rules in force at the time have been applied rigorously to all the available information.

An appropriate decision is one in which the decision-maker, in light of the eventual result, does not regret the selected course of action.

A decision to fund a project is based on some rules that simply show which of two options is consistent with the information provided in the business case: "fund/don't fund". If a decision was consistent with the organisation's decision rules, then it must be judged as "correct" (regardless of how the project subsequently turned out). So, at the end of a project, we could find ourselves in any of the four situations shown in Table 8.6.

Table 8.6 Possible conclusions that might be drawn from assessment of a project

Correctness of the original funding decision	Appropriateness of the original funding decision	
	Appropriate decision (investment was acceptable)	Inappropriate decision (investment was unacceptable)
Correct decision	"Just rewards"	"Unfortunate result"
Incorrect decision	"Lucky strike"	"Roosting chickens"

The actual values used in evaluation of investment success must cover all those effects attributable to the funding decision. The case study presented in Box 8.4 illustrates such a scenario.

Illustrative example

Box 8.4: The LAF Business Process Improvement Initiative: Fortuitous Outcomes Result in an Investment Success

A project undertaken by a large loss adjusting firm (identified here as "LAF Inc.") illustrates the way that fortuitous desirable outcomes should be handled in tests of investment success. Loss adjusting is the profession involved in the assessment of claims on insurance policies. The processes of loss adjusting are labour-intensive, using highly paid specialists. For some time, the loss adjusting profession has been coming under pressure from underwriters to reduce fees, and so profitability has been under threat. A few years ago, LAF responded by undertaking a significant project "Phoenix" to re-engineer its core business processes with two target outcomes: reduced operating costs (benefiting the firm itself) and reduced claim processing times (benefiting the underwriters, and their clients, the insured). A business case for "Phoenix" was accepted by the CEO, and the project eventually completed (significantly over budget and over time). Furthermore, the levels of reduction in both operating costs and claims processing time fell significantly short of their targets in the business case.

Near the end of the exercise, (in an unrelated move) the major insurance companies went out to tender for their in-house loss adjusting services. LAF's bid for this work was successful, largely because of the emergence (at that point) of fully documented, pilot-tested processes as output from project Phoenix.

The fortuitous eventual outcome "Increased revenue" was, of course, not included in the business case that had been accepted by the CEO, but was recognised in the final evaluation.

Despite the fact that project Phoenix was declared a project management failure and a project ownership failure, because the CEO would have accepted a business case that included the (fortuitous) increase in revenue, the project was judged as an investment success.

8.5.3 Qualifying Judgements About Investment Success

When it is directed at investment success, a regression test is triggered by the securing of target outcomes. The conditions under which target outcomes from a project are considered secured require predictions about the future flows of certain variables that then appear in the achieved business case. By their very nature, predictions involve uncertainty and so we have to make judgements about success conditional on the accuracy of the predictions involved. In other words, a regression test is, in general, carried out in the face of residual risk. This implies in turn that, in general, the declaration that a project is closed (and that its target outcomes have been secured) must be expressed in terms of a level of confidence (in the statistical sense).

The later a regression test is carried out, the larger will be the influence of actual data (and the smaller will be the effects of forecasts). Although this seems to suggest that the test of project success should be held off for as long as possible, such a decision would delay the date of project closure and, accordingly is not recommended.

8.6 Comparing the Three Tests of Success

8.6.1 Variables Used in the Different Tests

The three tests of project success (management, ownership and investment) can be summarised in the following way:

- All three use a form of regression testing in which trade-offs are permitted amongst criteria.
- All recognise a criterion called "cost", but each is calculated in a different way. Tests of project management success involve only project costs (they ignore any future operating costs), while ownership and investment success take both components into account.
- All recognise "time", but again, the views are different. Tests of project management success focus on the date by which all outputs were delivered. Project ownership success is concerned with the intended values and achievement dates of the variables that determine "worth" (as they appear in the business case). Invest-

ment success is concerned with (what the funder judges to be) acceptable values of the variables that determine worth at the time of the test.
- Investment and ownership success also include desirable outcomes that are not used in project management success.
- Two criteria associated with outputs are also accepted in the test of project management success. Not only must all outputs that are identified in the project scoping statement be delivered—but each must meet its specified level of quality. Outputs are not recognised explicitly in project ownership and investment success (although they have an indirect impact on both tests arising from their utilisation by project customers).

The three tests of success allow for eight combinations of success and failure—of which only six are valid. Despite the independence of the tests themselves, it would normally be expected that their results would be correlated. In other words, a project failing a test of project management success is more likely to fail (later) tests of ownership and investment success. Because it is difficult to obtain any information about the detail of original business cases, it is also very difficult to make judgements about ownership and investment success for well-known initiatives.

Illustrative example

Box 8.5: The Sydney Cross-City Tunnel—Application of the Three Tests of Success

Chapter 2 used the example of the cross-city tunnel, built to relieve congestion, pollution and noise in the CBD of Sydney, NSW Australia. Publicly available information seems to confirm that the infrastructure was delivered as specified, ahead of time and under budget. Furthermore, assume that there were no detrimental outcomes (Note that this assumption does not imply that there were no undesirable outcomes at all—some undesirable outcomes may have been identified earlier and accepted as part of the business case).

A target outcome was set as 90,000 vehicles using the tunnel each day. The assumption underlying this outcome is that the bulk of tunnel users would otherwise have made the same trip by driving on surface streets. Operational analysis, however, revealed that fewer than 30,000 vehicles per day have been taken off surface streets. Furthermore, the tunnel had to be sold because the operators became financially insolvent.

One could confidently judge the exercise as project management success because it met all four performance criteria laid down in the "steel tetrahedron"—scope/quality, timeframe, cost and detrimental outcomes.

It would appear appropriate, however, to judge the exercise as a project investment failure because we can reasonably infer from all that has happened since that the project would not have been approved by the state government if such an outcome had been known at the time of the original funding decision. What can be said about ownership success? Figure 8.5 shows clearly that investment failure implies ownership failure.

8.6.2 Valid Combinations of Judgements About Project Success

Figure 8.6 shows the eight combinations of judgements about the success of a project. We have eight combinations because there are three tests of success (related to project management, ownership and the investment, respectively), each one can result in either success or failure. Out of these eight options, only six are valid, because if a project has been judged an ownership success then it is automatically an investment success (but not vice versa). The figure is followed by examples of well-known projects illustrating the six valid combinations of project management, project and investment success.

L, P: Invalid combinations of success judgement (because ownership success always implies investment success).

J: US Aviation Automation System. A project to develop the (US) Federal Aviation Authority's Advanced Automation System (AAS) started in the early 1980s. AAS was intended to consolidate a large number of disparate items of air traffic control

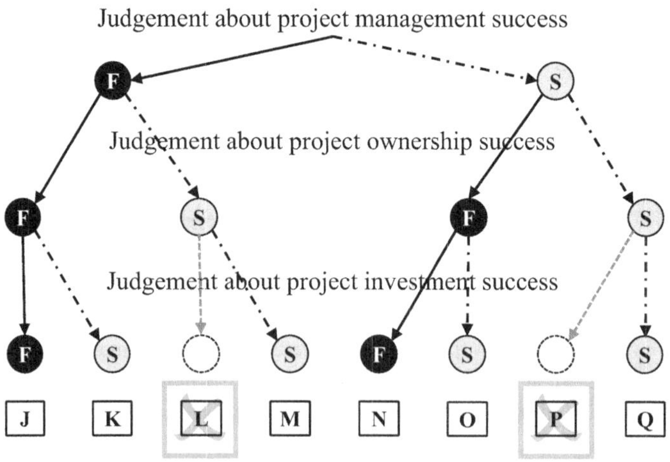

Fig. 8.6 The eight possible combinations that emerge from the three judgements about a project's success (s = success; f = failure)

infrastructure. The venture was terminated in the mid-1990s having exceeded its budget and timeframes significantly without delivering committed outputs from the original scope. This prevented realisation of any benefits and so the exercise was a failure in management, ownership and investment terms (GAO 1994).

K: Hubble Space Telescope. The telescope was produced, delivered and implemented with a faulty mirror. That, together with reported budget and timeframe overruns, suggest that the original venture was a project management (and probably a project ownership) failure. By way of contrast, the general view of the astronomy community (including NASA itself) is that the images generated by the instrument have been of inestimable scientific value, and thus the investment represented by the project was an investment success (Dunar and Waring 1999).

M: Boeing's 787 Dreamliner. The project was undertaken to respond to the growing demand for next generation, advanced and highly efficient aircrafts. The completion of the project was delayed by almost 40 months and it costs more than twice the original estimate. Yet, the 787 Dreamliner created a well-desired plane, which will continue to produce extensive financial business results (Shenhar and Holzmann 2017), hence can be considered as both project ownership and investment success.

N: Sydney Cross-City Tunnel. Details of the Cross-City Tunnel provided in Box 8.5 make it clear that this project is an example of a project management success and investment failure. It can also be reasonably inferred that it was also a project ownership failure (Zwikael and Smyrk 2012).

O: Los Angeles Metro's Red Line. The first 4.4 miles of the Los Angeles Metro, between Downtown and North Hollywood were opened to the public eight months ahead of schedule, with no budget overruns in January 1993 (Stapleton 1993). However, while expecting to see a million passengers during the first year of operation, the actual number was less than 60,000. Although eventually the remaining phases of the project were dropped, this project is classified as partial business success (Shenhar and Holzmann 2017), or, in our terminology, ownership failure but investment success.

Q: Moon landing. There is possibly no more memorable or eloquently expressed statement of timeframe for a project than that announced by President Kennedy in 1961 when proposing (what was to become) the Apollo 11 mission: "First, I believe that this nation should commit itself to achieving the goal, before this decade is out, of landing a man on the moon and returning him safely to the earth". Although available budget information is rather dated and fragmentary, this appears to have been not only a project management success, but an outstanding project ownership and investment success.

8.7 Critical Success Processes (CSP)

The previous sections introduced the way we gauge success. In order to assist managers with achieving success, this section introduces "Critical Success Processes"—an approach that seeks to identify those processes most important to success across a variety of project contexts. This approach is a more detailed and practical extension to the renowned concept of "Critical Success Factors", summarised in the next section.

8.7.1 Critical Success Factors (CSF)

To improve rates of project success, researchers have explored the main determinants of project success. These are usually called Critical Success Factors (CSFs).

CSFs are those areas that if addressed satisfactorily, will contribute significantly to organisational success. Daniel (1961) was the first to introduce the concept of CSFs in the general management area. Since then, the use of CSFs has become widespread in many areas, for example, strategic planning, knowledge management and quality management. The first application of CSFs in the project management arena was made by Rubin and Seeling (1967), who investigated the impact of various factors on project management success. They found that the size of previous projects to which project managers had been exposed was significantly correlated to success. The most cited CSF study has been conducted by Pinto and Slevin (1989). In their research, 418 project managers were requested to evaluate the importance of different factors relating to project success. The study identified the following ten CSFs:

1. **Project mission**. Initial clarity of goals and general directions.
2. **Top management support**. Willingness of top management to provide the necessary resources and authority/power for project success.
3. **Project schedule/plan**. A detailed specification of the individual action steps required for project implementation.
4. **Client consultation**. Communication, consultation and active listening to all impacted parties.
5. **Personnel**. Recruitment, selection and training of the necessary personnel for the project team.
6. **Technical tasks**. Availability of the required technology and expertise to accomplish the specific technical action steps.
7. **Client acceptance**. The act of "selling" the final project to its ultimate intended users.
8. **Monitoring and feedback**. Timely provision of comprehensive control information at each stage in the implementation process.
9. **Communication**. The provision of an appropriate network and necessary data to all key actors in the project implementation.
10. **Troubleshooting**. Ability to handle unexpected crises and deviations from plan.

Other studies, in different cultures, industries and project typologies, have also developed CSF lists, for example, Johnson et al. (2001). An analysis of these studies reveals that most of these lists share the following factors (Zwikael and Globerson 2006):

1. **Top management support**. Offered by the senior management of the performing organisation to properly support project management processes.
2. **Project plan**. A baseline document that provides all the information required to make a reliable decision about approving a start to work on a project and track a project's progress.
3. **Personnel recruitment**. Selection of the appropriate personnel for the project team.
4. **Monitoring and feedback**. Collection and analysis of information during project execution used to make decision with the way the project is managed.
5. **Customer involvement**. Active participation of the project owner and selected customers in the project management process.
6. **Project requirement and objectives**. Clear identification and definition of outcomes and outputs.
7. **Adequate spending**. Rigid plan and control of project expenses.
8. **Technical tasks**. Excellency in technical aspects of the development of the new output.
9. **Communication**. The provision of an appropriate network and necessary data to all key actors in the project implementation.
10. **Project strategy**. A high-level plan that focuses on how to achieve project's objectives.

8.7.2 The Need for an Alternative Critical Approach

The concept of Critical Success Factors (CSFs) has made an enormous contribution to the project management area. It allows project managers to focus widely and effectively on the most important areas to improve project management success. The identification of CSFs has been a significant step towards recognising the importance of core project management areas, such as project planning, top management support and customer involvement. As a result, a number of effective tools, techniques and models have been developed in recent years to support these critical areas.

Some criticism has been directed at the CSF approach (Fortune and White 2006; Zwikael and Globerson 2006). The critiques, while accepting the importance of CSFs, also highlight their limitations. The major identified weaknesses include:

1. **CSFs no longer provide additional practical knowledge for project managers**. Project management CSFs have been studied for more than 50 years

(Rubin and Seeling 1967). The importance of CSFs is acknowledged by most project managers who already focus on critical areas. Moreover, most project management tools and software packages also support these recognised areas such as project planning and project control. However, CSFs do not advance our understanding of project management success. They also ignore the history and context of individual projects. Although CSFs may increase the general knowledge of project managers, they are not specific enough to support better decision-making. For example, the importance of "top management support" is generally accepted, but managers still have little by way of guides and practices regarding how to obtain this type of support.

2. **No one set of CSFs suits all projects**. Contingency theory (Donaldson 2001) suggests the deployment of choice should be dependent upon variables of firms and environmental conditions. In the project environment, this means that different projects require exclusive managerial focus, for example, for projects with high versus low levels of risk, project executed in different industries and cultures, as well as for different levels of project pace, novelty, complexity and technology (Shenhar and Dvir 2007). A single list of CSFs is unlikely to fit all project scenarios. If CSFs are to be implemented in a particular project, they should recognise its unique features.
3. **CSFs are not sensitive to different project phases**. The factors that demand special attention from a project manager change throughout the project's life because different project phases have different characteristics. In addition, project managers find that the nature of their work changes from phase to phase. For example, planning requires organisation and analysis, while execution demands highly developed people skills. Most CSFs studies present lists of factors that are both static and generic across all project phases. Therefore, unique CSFs should apply for each project phase. For example, it is generally accepted that a focus on project strategy at the earlier phases of a project is important. Pinto and Slevin (1990) suggested different CSFs that should be applied across the project's life. Table 8.7 analyses the relative importance of the ten most recognised CSFs (Zwikael and Globerson 2006) across project phases.

8.7.3 The Critical Success Processes (CSP) Model

Because traditional CSFs are so general and do not include guidance for project managers' decision-making, an alternative approach is required. Some researchers have accepted the need for a more detailed list of critical project management processes. While the Critical Success Processes (CSP) model relies on the theoretical foundation of the CSF concept, at the same time it represents a more detailed approach. Instead of providing managers a general list of critical factors, it identifies specific project management processes that most effectively contribute to success in different

Table 8.7 The relative importance of critical success factors in various project phases

Critical success factors	Initiation	Planning	Execution	Outcomes realisation
Top management support	**	**	***	**
Project plan	**	***	*	
Personnel recruitment	**	**	***	
Monitoring and feedback			***	
Customer involvement	***	***	***	***
Project requirement and objectives	***	***	**	*
Adequate spending			***	
Technical tasks	*	*	***	*
Communication	**	**	***	**
Project strategy	***	**	*	

*** Extreme importance
** High importance
* Medium importance

project scenarios. Executives can ensure that these critical processes are supported by appropriate procedures, templates and tools in the organisation. Project managers can then enhance their effectiveness by devoting more attention, effort and time to these critical processes.

The concept of CSP is more focused, exact and practical than the traditional CSF approach because it recognises the peculiar requirements of different project scenarios. A research model has been developed in which the dependent variables are judgements about the various forms of project success and the independent variables take the form of project management processes. This list of processes has been drawn from literature covering fields as diverse as project management, organisational support, organisational planning, organisational control and maturity modelling. For more information about this model, see Zwikael and Globerson (2006).

Finally, there is a growing recognition that different types of projects require dissimilar approaches to their management (Shenhar and Dvir 2007). Since CSPs are expected to vary across project scenarios, some potential moderating variables are also included in the model. This means that the impact of various project management processes on project success can vary across project scenarios. This also raises the need to identify particular CSPs for different project scenarios, as suggested in the following model. Some variables included in Fig. 8.7 may suggest that various project management and project benefit management processes have different effect on the various levels of project success across projects. These moderating variables are level of project risk, industry, culture, project pace, novelty, complexity and technology (Zwikael 2008).

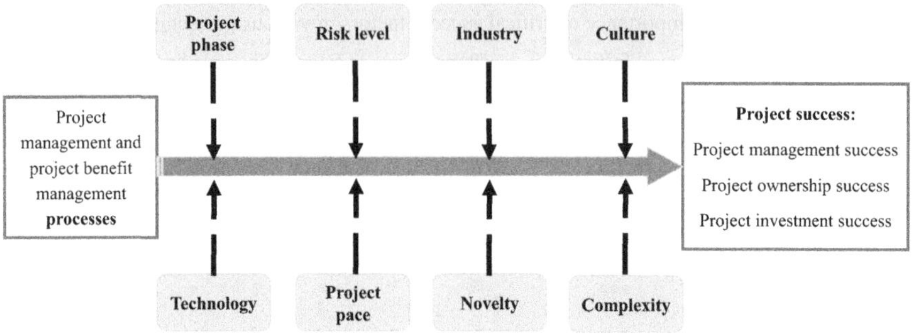

Fig. 8.7 The Critical Success Processes (CSPs) research model

Some initial results of this approach have recently been published, focusing on achieving project management success. While the identification of project activities was found to be the most critical project management planning process (Zwikael and Globerson 2006), it has also been found that the level of contribution of different processes to project success varies across sectors. Quality management is the most important process in the service sector (Zwikael and Globerson 2007), scoping in the information technology sector (Zwikael 2008) and cost management in the construction industry (Zwikael 2009). Similarly, Box 8.6 demonstrates the most critical top management support processes in different industries.

Practicalities

Box 8.6: CSPs across Project Contexts—The Importance of Different Top Management Support Processes in Various Industries

We use the CSP model to identify project management processes that appear to be particularly relevant in different project scenarios and phases. Later chapters describe the most critical processes that can be used for each of four project phases: initiation, planning, execution and outcomes realisation.

Top management support is a recognised CSF. This alone, however, does not provide executives with specific tools to support project managers effectively. A study based on the model (Zwikael 2008) shows the importance of different top management support processes for projects executed in various industries. Table 8.8 presents the most critical top management support processes in each industry.

Table 8.8 Critical success top management support processes in different industries

Top management support process	Industry
Existence of project procedures	Construction, government
Appropriate project manager assignment	Software, engineering, government
Refreshing project procedures	Software, construction
Communication between the project manager and the organisation	Software, production
Existence of project success measures	Software
Supportive project organisational structure	Government, construction, software
Existence of interactive interdepartmental project groups	Software, construction, communications, government
Organisational project resource planning	Software, engineering, government
Organisational project quality management	Construction
Ongoing project management training programmes	Construction, government
Project management office involvement	Services, software
Use of standard project management software	Software
Use of organisational projects data warehouse	Production

According to Table 8.8, the most significant impact on project management success in each industry arises from different top management support processes. The results of this study confirm that exclusive CSPs exist in different industries. As a result, one should tailor project management approaches to the industry where the project takes place, instead of implementing generic best practices.

8.8 Tests of Success as Special Case of the Three-Layered Model

The three-layered model of project success is consistent with recent discussion of project success dimensions (Lechler and Dvir 2010), where there is agreement on the importance of long-term organisational benefits achieved due to the project. Amongst the success criteria suggested by these writers are included such end effects as the business impact on the organisation, the opening of new business opportunities for the future, the ability to use the customer's name as a reference, benefits to the community, or any other target outcome stated by the funder in the business case. In

Table 8.9 Project success dimensions in Lechler and Dvir (2010) mapped to those introduced here

Project success dimensions (Lechler and Dvir 2010)	Three-layered model of project success (to which the proposed dimension belongs)
The project had come in on schedule	Project management success (timeframes)
The project had come in on budget	Project management success (cost)
The project met all technical specifications	Project management success (scope)
The results of this project represent an improvement in client (funding organisation) performance	Project ownership success/investment success (desirable outcomes)
The project is used by its intended clients (customers)	Project ownership success/investment success (output utilisation leading to desirable outcomes)
Important clients, directly affected by the project, make use of it	Project ownership success/investment success (output utilisation leading to desirable outcomes)
Clients (customers) using this project will experience more effective decision-making and/or improved performance	Project ownership success/investment success (desirable outcomes)
The project has a positive impact on those who make use of it	Project ownership success/investment success (desirable outcomes)
The clients (funders) were satisfied with the process by which this project was completed	Project ownership success/investment success (desirable outcomes)
The clients (funders) are satisfied with the results of the project	Investment success
The project was an economic success for the organisation that completed it	Investment success
All things considered, the project is a success	Investment success

other words, all other leading project success studies include success measures that are special cases of the generic framework presented in this book.

As part of an exercise to validate the tests of project success, we have analysed the alignment between the determinants of the proposed project success tests (project management, ownership and investment) with those appearing recently in the project management literature. For example, Lechler and Dvir (2010) measure project success using the satisfaction of the project funder ("customer" in the original terminology of their paper). It would appear from the following table that this approach is a special case of the three-layered model of project success.

In Table 8.9, the terminology related to the three-layered model has been added in brackets to the original success measure for easy comparison.

The analysis in Table 8.9 suggests that the dimensions appearing in recent literature can be viewed as special cases of those proposed here. The large number of different dimensions that can all be considered as specific instances of "desirable outcomes" (only some are included in the models discussed in the literature), is particularly noteworthy. In addition, the work of the other scholars does not take

Table 8.10 Project success dimensions in Kerzner (2013) mapped to those introduced here

Project success dimensions (Kerzner 2013)	Three-layered model of project success (to which the proposed dimension belongs)
Within the allocated time period	Project management success (timeframes)
Within budgeted cost	Project management success (cost)
At the proper performance or specification level	Project management success (scope)
With acceptance by the customer (funder)	Project management success (quality)
With minimum or mutually agreed upon scope changes	Project management success (scope)
Without disturbing the main workflow of the organisation	Project ownership success/investment success (undesirable outcome)
Without changing the corporate culture	Project ownership success/investment success (undesirable outcome)

into account the possibility of a reduction in ownership success due to undesirable outcomes.

A second comparison is based on success dimensions discussed in Kerzner (2013) is provided as Table 8.10.

Table 8.10 shows that all of Kerzner's seven success dimensions are special cases of the success measures described earlier. From this analysis, we conclude that our framework comfortably accommodates all the success dimensions proposed by others.

Summary

In this chapter, we have assembled a general framework for handing the concept of success as it relates to projects. This involves distinguishing project management success, project ownership success and investments success. In light of this framework, we have examined the conventional approach to gauging success in projects and found it wanting, not only as a test of project success and investment success, but also (and perhaps surprisingly) as a test of project management success. By exposing the underlying weaknesses in the accepted approach, we have assembled tests that derive directly from the underlying theory of the ITO model. Part A of the book has provided us with some foundation concepts that are common to the management of all four of a project's four global phases. In Part B, we discuss the conduct of each of the four project global phases—initiation, planning, execution and outcomes realisation.

Part II
Leading a Project

Chapter 9
Initiating a Project

Abstract When a new project is being considered, the funder needs to know enough about the venture to make a reliable investment decision. The details necessary to support such a decision are presented formally as a business case. It is during initiation (the first of the four global phases spanning the life of a project) that the business case is assembled and appraised. This chapter explores initiation in some depth—including discussion about the processes involved, the roles filled by key players, the tools that are used and the structure of the business case itself.

9.1 Overview of Initiation

When a new project is being considered, the funder needs to know enough about the venture to make a reliable in-principle investment decision. The details necessary to support such a decision are presented formally as a business case. It is during initiation (the first of the four global phases spanning the life of a project) that the business case is assembled—and the project then considered for progress to the next global phase (planning).

This chapter explores initiation from the perspective of the funder and other key players—with the objective of helping them understand how the business case is assembled, interpreted and approved.

9.1.1 Initiation in Outline

The initiation phase starts with an idea for a new project and finishes with a decision made on a business case presented to a potential funder. Initiation is made up of four sequential sub-processes (see Fig. 9.1):

1. Identification
2. Definition
3. Analysis
4. Packaging

O. Zwikael and J. R. Smyrk, *Project Management*,
https://doi.org/10.1007/978-3-030-03174-9_9

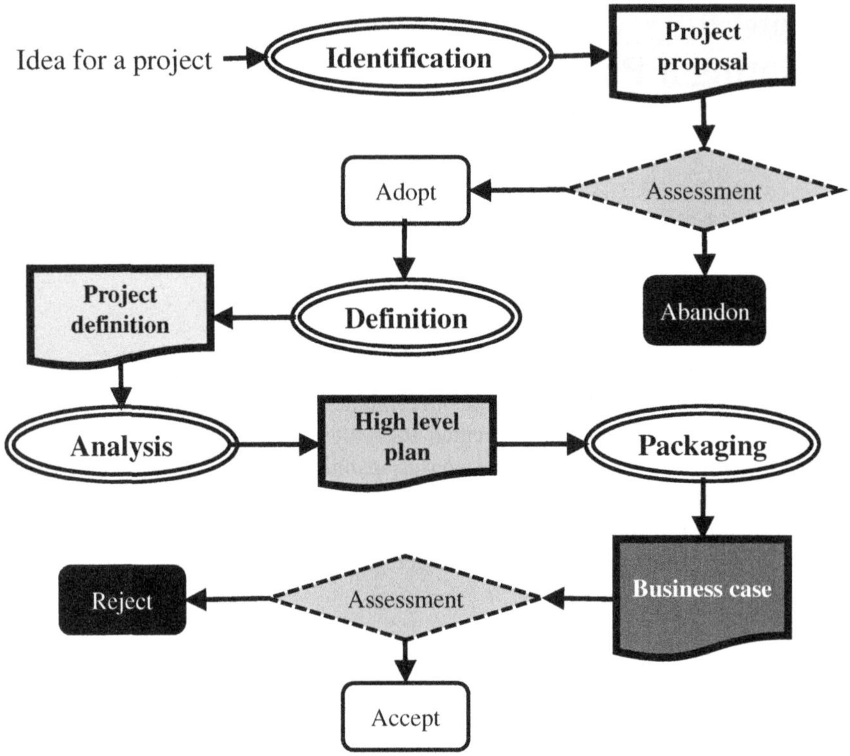

Fig. 9.1 The project initiation process

It should be noted that these sub-processes are *sequential*. The business case cannot be packaged until the analysis is complete. Analysis can only be carried out when the project definition is available. A definition cannot be assembled until the project has been identified. It is possible, however, that when working on one sub-process it is revealed that further work is required on an earlier one.

The initiation of a project is triggered by either of two events: the funder approves a request from someone to prepare a proposal, or the funder instructs someone to prepare a proposal. A project proposal is the output from *identification* of a new potential project (the first of the sub-processes in initiation). If the proposal is accepted, then initiation progresses to the second sub-process—*definition*. It is here that a number of core parameters are established and confirmed with the funder as a *project definition*. The definition sets the overall "shape" of the project—including lists of target outcomes, committed outputs and downstream processes. At this point, the funder and the champion share a common view of the project that is to be the subject of the business case.

During the analysis that follows definition, various aspects of the project are considered, including a high-level script for the eventual third global phase (execution), stakeholders, risks, issues and governance arrangements. The preliminary script is then used to derive tentative estimates of duration and cost. Together with the project definition, and some background to the initiative, the results of the analysis are incorporated into a business case. Initiation ends with a decision by the funder to either accept or reject the business case as the foundation for the next global phase—planning.

9.1.2 *Key Players in Initiation*

Because it is not always appropriate for the person who came up with the original idea for the project to lead business case development, this activity will often be driven by someone else, who is identified as the project "champion". The prospective project's funder appoints the project champion. Because the person appointed as project champion will typically lack the technical skills and time required to actually write the business case, someone else is appointed to assist with that work. Initiation involves, therefore, three key players: the funder, a project champion, and (if needed) someone who undertakes most of the detailed activity (These appointees are often, in effect, the project-owner-designate and project-manager-designate respectively).

The project champion (who may already have been confirmed as the project owner) leads initiation. The funder plays a relatively passive role—but should be kept up to date in regularly briefings by the project champion. In general, champions (as prospective project owners) are drawn from the ranks of senior executives in the funding organisation.

9.1.3 *Key Issues in Initiation*

The business case is the first of two baseline documents—the other being the project plan. A baseline document is one on which all substantive subsequent activity and decisions are based. To understand the methodology behind the assembly of the business case it is necessary to recall that, *during project execution*, the ITO model progresses *from left to right*. Resources are acquired, then deployed. Work is undertaken to produce various outputs—which are then delivered to the project's customers. The customers, in turn, utilise these outputs as they participate in particular downstream operational processes. Target outcomes eventually emerge as changes in specific operational metrics attached to those downstream processes.

Assembly of the business case, however, requires that the ITO model be applied *right to left*. The logic of this flow rests on a number of observations. A business case includes (amongst other things) details about resources and timeframes. Before estimates of cost and duration can be produced, it is necessary to understand what sort of below-the-line work is involved. Because below-the-line work is undertaken exclusively to produce the project's outputs, to describe that work we must have a definitive list of committed outputs. Eventually, the committed outputs from a project must be capable of being utilised (and hence generating target outcomes). That capability is expressed as a list of necessary fitness-for-purpose features.

The lists of committed outputs and target outcomes must be set, validated and agreed before anything can be said about the project's other core parameters. Getting these lists right is known as the scoping problem—which can be expressed as a two-part question: "Given a list of *target outcomes*, how do we know if: (A) the list of committed outputs and (B) the definitions of all outputs, are 'right'?". Because historically there has been no obvious way of obtaining a clear, reliable and, more importantly, *reproducible* answer to this question, the scoping problem has proven to be almost intractable. Below (Sects. 9.3.1 and 9.3.2) we propose an analytical technique to resolve it in which a project is scoped at *two* levels: the first considers the list of outputs to be delivered by the project, the second addresses the fitness-for-purpose features that will define each of those committed output.

The business case includes estimates of certain parameters which will be used by the project funder to gauge the expected worth of the exercise—and hence make a reliable decision about investing in the exercise and progressing to the second global phase (planning). When estimates are provided to support decision-making processes, they should be qualified with an indication of their reliability. Of the various available techniques to derive a measure of reliability, one (which is widely in the project management discipline) is to provide three figures: lower bound, most likely, upper bound. The *most likely* serves as the notional estimate.

9.1.4 The Relationship Between Initiation and Planning

To fully understand project initiation, it is important to be aware of some subtle relationships between the business case and the (subsequent) project plan. As is discussed below, a business case provides the funder with three sorts of detail: a *project definition* (based on the lists of target outcomes, committed outputs and downstream processes), a *high-level plan* (including schedules and resource estimates) and *contextual information* (related to such things as the background to the proposal, stakeholders and risks). By way of contrast, planning is primarily concerned with the systematic analysis of the project's definition (and subsequent reworking) of the high-level plan. During that work, additional contextual information also becomes available with the result that during planning much of the original business case is, effectively, overwritten with additional detail (for example, the work to be done in execution), or revised figures (such as cost and time estimates if different from the original estimates). The project definition (which is the core of the business case), is accepted as a given and so remains unchanged by the work undertaken during planning (If, for some reason, it is discovered during planning that the definition is wrong, then the business case must be reconsidered for approval by the funder before planning can continue).

A project's attractiveness based on its business case may differ from that based on its project plan. This situation will arise if planning reveals that the estimates of cost, duration and riskiness contained in the business case require revision. In view of all this, the business case should establish bounds on these parameters, outside of which the project owner will require the approval of the funder.

The tools and techniques used during initiation to assemble the high-level plan contained in the business case and, later, during planning to develop the (much more substantial) project plan are fundamentally the same. The difference lies solely in the depth of the analysis which they support. For this reason, in what follows, we introduce them with only as much detail as is necessary to understand how they fit into the initiation process. Each is later covered more comprehensively in Chap. 10.

Planning intentionally overwrites certain parts of the business case with more complete (and presumably, more reliable) information. For example, the high-level workplan section of the business case is *completely replaced* with extremely detailed versions in the project plan). What is the reason for all the reworking that takes place in planning? The ITO model requires that a reliable script be available before execution can be undertaken, however, the high-level script produced during initiation is inadequate for this purpose. So why do we bother with a high-level script in the business case at all? Why not just limit initiation to the project's definition and the provision of contextual information? The answer lies with the vastly different amounts of work required to produce the high-level plan in the business case and the comprehensive version in the project plan. Assume for a particular project, that the funder would be prepared to invest no more than $M5, however, a high-level plan (if produced as part of a business case) would reveal that costs lie in the (wide) range $M6–$M8. Without any high-level planning information in the business case, the funder would be forced into incurring the significant costs (and delay) of project planning before deciding to abandon the exercise. A business case, on the other hand, allows the funder to avoid the costs of planning before making such a decision (Clearly the reliability of all estimates provided in the business case must be qualified at the time of their tabling).

In Sect. 9.4.3 (The impact of planning on the business case) we explain, element-by-element, which parts of the business case are expanded by planning process.

Illustrative example

Box 9.1: Project SmartDeal: A business process improvement initiative
The illustrations that are provided of various concepts throughout this chapter all relate to the following SmartDeal case study.

The manufacturer of a range of kitchenware products (PKW Inc.) has undertaken a study into its procurement practices. This has revealed the following major shortcomings compared with their main competitors:

- High levels of working capital.
- Unacceptably long settlement times of supplier invoices.

- Long durations (lead times) for procurement.
- High levels of errors in orders placed on suppliers.

A project (named "SmartDeal") was considered to address these issues.

Early work on the business case suggested the following elements of project scope:

A. Target outcomes:
   - Reduced settlement times for supplier invoices.
   - Reduced lead time for procurement.

B. Committed outputs
   - A panel of preferred suppliers.
   - A new structure for the national purchasing office.
   - Enabling information systems and infrastructure.
   - A programme of professional accreditation for procurement staff.

C. Downstream processes:
   PKW view the overall procurement process as made up of two (related but independent) sub-processes:
   - Internal procurement: whereby an operational or functional area requests the supply of particular resources from the national purchasing unit.
   - External procurement: where, in response to an internal procurement request, the national purchasing unit places and manages an order on a supplier.

9.2 Project Identification

Identification is the first sub-process in the project initiation phase. A prospective project is identified in a project proposal. A project proposal can emerge as a response to ideas from many different quarters such as a prospective funder, a business unit manager, a services department, an employee, an external consultant or even a supplier. For example, the marketing manager may suggest developing a new product to meet a specific gap in the market as an opportunity to increase market share. Other projects may start because of a constraint, for example, new legislation requiring a reduction of 20% in the greenhouse gas emissions produced by an organisation. Here, the proposal to prepare a business case could come from the organisation's legal advisor. Whoever is to table the project proposal fills the role of project champion, and so, the key players in the project identification are the funder and a project champion.

If a prospective project is taken through to the end of initiation, a business case is tabled and considered by the funder for an investment decision. A firm commitment to start execution is conditional on the approval of the project plan.

Assembly of a project proposal requires no formalised identification methodology—the proposal can be attacked in any way that suits both key players. A generic template for the project proposal is provided in Box 9.2.

Template

Box 9.2: The Project Proposal

1. **Name of project**: A short title (i.e. a "nickname") that clearly identifies the project.
2. **Objective**: A short statement that answers the question "Why should this project be funded"?
3. **Key players**: Identify the funder and the project champion.
4. **Background**: How did this idea for a project arise and what is known about it? What problem does it aim so resolve, the opportunity to explore or constraint to deal with?
5. **Next steps**: If the proposal is adopted—in particular, who will be involved in the preparation of the business case, and how long will it take to assemble?

Some observations about this template

1. **Name of project**. A project can be called anything—as long as it clearly identifies the project and distinguishes it from others in the funder's portfolio of projects. Care must be exercised in the adoption of a name which pre-empts the project's scope. For example, such names as "The XYZ *system* project" is undesirable because it implies a single output (a "system").
   As a practical matter, the name of a project should be short—so that it is readily accepted as part of the organisation's lexicon.
2. **The objective** should have four characteristics:
   - It is short.
   - It is framed in outcome terms.
   - It clearly carries the intent of the project.
   - It begins with "The objective of this project is to …"

If a project has only one target outcome, then that may well serve as the statement of objective, for example—"The objective of this project is to reduce customer service costs" (Target outcome = Reduced customer service costs).

If a project has multiple target outcomes, then the statement of objective must be general enough to encompass all of them. "The objective of this project is to make our community a safer place for children" (Target outcomes: reduced incidence of child abuse, reduced recidivism amongst sex offenders).

It is not required that a statement of objective be expressed in a way that it can be measured (The project's eventual target outcomes will serve that purpose).

3. **Key players**.
   - The funder is a person with the authority to spend the money.
   - The champion is the person who will guide the assembly of the business case and later, present it to the funder.
4. **Background**.
   - What has led to this project being proposed? Typically, this will take the form of an opportunity, problem, or a constraint.
   - Why is the project important?
   - What is an example of an outcome that *might* be targeted?
   - What are two or three of the more prominent outputs that *might* be produced during project execution?
5. **Next steps**.
   - What will happen next if this proposal is adopted?
   - Who will be involved in initiation, and roughly how long will it take?

When adopting the project proposal, the funder should then decide who will lead the next stages of initiation and develop a full business case. Two immediate options can be considered

- Appointing a project champion to guide initiation. This option foreshadows a decision (to be made when approving a start to planning) about who will be appointed as project owner.
- Appointing a project owner to do this. In this case, the project owner fills the role of project champion.

9.3 Project Definition

Definition is the second sub-process in the project initiation phase. Once the proposal has been adopted, work can start on defining the project. At the end of this activity, a project definition will have been confirmed and accepted by all key players as the basis for the following *analysis* activity (There is no decision point at the end of definition—it simply continues until the definition is suitable for analysis).

9.3.1 A Project Is Scoped at Two Levels

The eventual business case establishes the scope of the project at each of two levels—in that it contains a list of committed outputs and, for each output, a definition (in the form of a list of necessary fitness-for-purpose features). The approach to the assembly of the business case discussed in what follows addresses the two levels sequentially. First, the project's committed outputs are identified, then each is defined.

Identification of a project's outputs is, in general, a creative/exploratory process—requiring a good understanding of the project context. The resulting list of committed outputs appears in an (initial) statement of scope which is later validated using some tools discussed in what follows. The statement of scope does not define those outputs (that is done later in initiation).

9.3.2 Overview of the Definition Activity

The project definition establishes the limits of the project—the core of which is represented by three clear and unambiguous lists: target outcomes, committed outputs and downstream processes. By setting scope in this way, the work and resources for the project are completely bounded. With scope locked down, any uncertainty about time and cost in the eventual business case can arise only from uncertainty in estimates of parameters. A carefully considered statement of scope provides a robust foundation for recognising, analysing and resolving issues of "scope creep" (the tendency for projects to grow in size over time).

Definition is a naturally iterative process involving ad hoc cycles of five activities

- Exploration of alternative project scopes.
- Establishment of a tentative statement of project scope.
- Validation of project scope.
- Definition of target outcomes and committed outputs.
- Packaging of the project definition.

It does not matter where the work begins—nor does it matter about the order in which the iterations progress.

A template for the project definition is provided in Box 9.3—with some supporting discussion of the items appearing there. The rest of this section provides a detailed description of the process that will be executed to populate this template. Because they are all approached iteratively, the order in which activities are undertaken does not normally align with the headings in the template.

Template

Box 9.3: The Project Definition

1. **Name of project**.
2. **A statement of project scope**. This has five elements:
   - Project objective.
   - The funder.
   - A list of target outcomes.
   - A list of committed outputs.
   - A list of the project's downstream processes.
3. **Target outcome definitions**.
4. **Validation of project scope:**
   - Overview of scope validation.
   - Outcomes coverage matrix (OCM).
   - Integrated utilisation maps (IUMs).
5. **Committed outputs definition**.
   For each committed output:
   - A list of *included* fitness-for-purpose features (FPFs).
   - A list of *excluded* fitness-for-purpose features.
6. **Scope boundaries:**
   - A list of selected excluded target outcomes.
   - A list of selected excluded outputs.
   - A list of selected excluded downstream processes.

Some observations about this template:

1. **Name of project**: This should either repeat the title adopted in the project proposal—or appear as a refined version of that.
2. **A statement of project scope**:
   - **Project objective**. This should either repeat the objective declared in the project proposal—or appear as a refined version of that.
   - **The funder**. The name of the entity (or entities) with the authority to approve the (eventual) business case and commit funds to the proposed project. Note that a project can have multiple funders.
   - **A simple list of target outcomes.** (Names only—all relevant details appear in the target outcomes definition table). Rarely will there be more than five, but the fewer the better.

- **A simple list of project outputs**. (Names only—relevant details will appear the subsequent committed outputs definitions). There will be as many outputs as necessary to support the achievement of the target outcomes.
- **A list of the project's downstream processes**. (Names and, where appropriate, a brief descriptor—bearing in mind that a comprehensive description appears later as part of the subsequently integrated utilisation map). Because a downstream is process is operational—it should be given the name used to identify it by those working in the associated operational environment, for example: "fulfil customer order", "procurement", "manufacture product" and "commute from suburb to city centre". Each downstream process should be classified as endogenous/exogenous. We remind the reader here that in Chap. 2 we classified downstream processes according to their origins:
  - *An endogenous downstream process* is produced as one of the project's outputs—and then imposed on the operational environment. For example, the procurement process in Project SmartDeal (Box 9.1).
  - *An exogenous output* is imposed on the operational environment from outside the project. For example, the channel navigation process in the Ripple Rock project (Box 2.8).

3. **Target outcome definitions**:
   This takes the form of a table with one column for each target outcome and seven rows—one for each of the attributes that define a target outcome.
4. **Validation of project scope**.
   The tools used to scope the project are the outcomes coverage matrix (OCM) and (a set of) integrated utilisation maps (IUMs).
5. **Committed outputs definition**.
   This takes the form of a two-level hierarchy
   - A list of committed outputs.
   - A list of the fitness-for-purpose features for each committed output—as revealed by the IUMs.
6. **Scope boundaries**
   Three lists help set the boundaries of the project by declaring, effectively, what is *not* in scope in terms of (excluded): target outcomes, outputs and downstream processes. The conditions under which an item qualifies for inclusion in these lists is discussed in detail below. Broadly, early in initiation, various stakeholders will be involved in exploratory discussions about the scope of the project. These discussions will result in numerous suggestions about the possible shape of the project—including tentative ideas for outcomes, outputs and downstream processes. So that expectations about the project can be managed, it is important to identify all those ideas which were considered, but not adopted.

9.3.3 *Setting the Scope of a Project*

The "scope" of a project conveys the idea of "things that are included and things that are not". The usefulness of such lists depends on what is meant by "things". One could, for example, attempt to define scope in terms of the activities that will be funded and those that will not. Because activities are undertaken to produce outputs, to decide if a particular activity is in or out of scope, one must know if the related output is required. Conclusion? The scope can be set meaningfully only with reference to a project's *outputs*.

In the following discussion, we address, in turn: the concept of scope, the way in which scope is declared and the different types of output that represent valid items for inclusion in a project's scope.

9.3.3.1 The Concept of Project Scope

A project is scoped when the following are identified (and listed): target outcomes, committed outputs and downstream processes. The identified list of *committed outputs* serves later as the foundation for the third sub-process of initiation—*analysis*. The focus of analysis is a tentative script for the (eventual) third global phase of the project—execution. In Sect. 9.3.3 below we discuss the way in which preliminary working versions of the three lists (target outcomes, committed outputs and downstream processes) are assembled. In Sect. 9.3.4 we explain how two scoping tools are used to validate working lists for inclusion in the project's scoping statement. The last subsections of 9.3 describe how target outcomes and committed outputs are defined.

Recall that target outcomes are generated when project customers utilise the outputs from a project during the execution of downstream processes. These are called *ITO outputs*—all of which must appear in the list of committed outputs. The list of committed outputs will, however, also include certain necessary outputs that play no part in the execution of downstream processes—such as those required for stakeholder management or risk mitigation. These *non-ITO outputs* are discussed in Sect. 9.3.3.3 below.

By convention, above-the-line outputs (such as the business case) are *not* included in a project's scope (because they are *always required for every project*). Intermediate outputs (such as architectural concept drawings and bracket assembly instructions) also do not appear is the statement of project scope for the same reasons that apply to above-the-line outputs.

A project can be correctly or incorrectly scoped. A correctly scoped project is one in which, given the target outcomes sought by the funder, the list of committed outputs

A. Includes all ITO outputs necessary for the successful execution of the project's downstream processes.
B. All necessary non-ITO outputs.
C. No redundant outputs of any kind.

A project can be incorrectly scoped in either of two ways: underscoping or overscoping. Underscoping is the situation that arises when some outputs from lists A and B are missing. Underscoping can arise from a variety of causes—especially a failure to establish a causal link between outputs and outcomes—resulting in critical outputs being missed. It may also occur because an underscoped project can appear superficially attractive to a funder (In general, as the list of outputs reduces, costs fall, and execution time is reduced). At the same time, an underscoped project cannot generate its target outcomes. Anecdotal evidence suggests that when projects get into trouble (usually by exceeding either their timeframes or their budgets), a common response is to remove outputs (descoping)—thereby leaving the project underscoped. If the relationship between target outcomes and outputs is not explicit or, (worse still) if target outcomes are not declared at all, then the impact of descoping on eventual project success may never be known.

Overscoping is the situation that arises when the list of committed outputs includes redundant items. Overscoping will pointlessly reduce the worth of a project (by inflating cost and duration without any compensating increase in benefits). Like underscoping, overscoping can arise when the linkages between outcomes and outputs are not established—resulting in outputs not being identified as redundant. It can also arise because of poor budgeting practices where an arbitrary pool of money for one project is seen as some sort of "bank" for the production of outputs that really belong to other projects.

9.3.3.2 The Statement of Project Scope

Scoping a project is led by the project champion (or the owner, if already appointed), but the actual work involved is undertaken by someone with the necessary project management skills. It begins with the identification of the funder—the entity with the discretionary authority to commit funds to the venture. This must be done first because only funders can decide if the eventual list of target outcomes makes sense. Scoping itself is now directed at the assembly of initial (usually tentative) lists of

- Target outcomes.
- Committed outputs.
- Downstream processes.

The early versions of these lists can be used as the basis of interviews with whoever the funder, the champion and the project-manager-designate believe can contribute to the definition of the project. Because those who are involved in the discussions do not necessarily use terms such as outcome, output and process as defined in this book, it is important to use whatever terms that the interviewee feels comfortable with. The results of these discussions will typically be a grab-bag of outcomes, outputs, business processes, key issues, downstream processes and other things. It is up to the project-manager-designate to catalogue and correctly classify these suggestions "offline" in the follow-up to each interview.

As the discussions progress, the three lists will grow in size—often suggesting a project with impossibly large scope. That is not an issue at this point because the mere fact that the interviewees are contributing to the discussion becomes an important form of engagement.

When it is believed that all critical input to these lists has been obtained, it is then necessary to cull them so that the list of candidate target outcomes is manageably small (typically no more than five—but often no more than two or three). Consider a target outcome which (for any reason) is to be removed from a project's scoping statement. Because the generation of that outcome requires that certain outputs get utilised, they, too, will appear in the statement of scope. Removal of an outcome allows any output which is unrelated to any other target outcome to be culled. Culling of *outputs* is achieved by applying four criteria to each of the candidate *target outcomes*. Any outcome that fails any criterion is rejected from further consideration. The criteria are

- **Importance**: Is this is an outcome which *the funder(s)* (alone) sees as worthy of investment?
- **Plausibility**: Can a plausible link be established between the project and the achievement of the candidate outcome? Recall that a target outcome must take the form of a metric for a downstream process. For example, "increased contribution margin" will often be suitable as adoption for certain types of projects, but "Increased annual profit (for the business)" will usually be unsuitable because corporate profit is influenced by so many factors outside the project. Looked at another way, "Increased annual profit (for the business)" is unlikely to serve as a useful metric for *any* operational process. A useful way of testing plausibility is to ask, "*If I wanted to achieve this particular outcome, is the proposed project an obvious way of doing that?*".
- **Lead time**: Is the lead time between the allocation of funds to the venture and the point at which investment success will be judged acceptable to the funder? Take a "Quit smoking" initiative. An outcome such as "Reduced smoking rates amongst teenagers" would be completely acceptable because we would expect to see such an effect within months of the end of project execution. "Reduced rates of lung cancer", would, on the other hand, be unacceptable because such an effect might not be observed for some decades. Although unacceptable as a *target outcome*, this would, however, be an excellent candidate for the *statement of project objective*: "The objective of this project is to reduce the incidence of lung cancer".
- **Measurability**: Can the outcome be measured within an acceptable timeframe and for an acceptable cost?

In addition, Zwikael et al. (2018) defined three dimensions of effective target outcomes: specificity (e.g. specific target values), attainability (e.g. the capacity to realise the target outcomes), and comprehensiveness (e.g. reflect the views of key stakeholders).

The application of these criteria will normally reduce the list of potential target outcomes significantly. If, however, it is still too long, then the funder could be asked to set a cut-off in a list of the candidates ranked by their importance. In light of the reduced list of candidate target outcomes, the working lists of outputs and downstream processes can usually be reduced "by eye". It does not matter if this culling is incomplete—the validation activity (discussed below) will allow appropriate additional changes to be made. We now have a tentative statement of project scope—which may now be validated (Sect. 9.3.4).

The list of target outcomes, the lists of committed outputs and downstream processes in the scoping statement will have been assembled intuitively, and, because of that, it may have redundant (or missing) outputs. Errors of that kind are not normally of concern at this point because validation will expose them.

9.3.3.3 Types of Outputs Appearing in the Statement of Project Scope

The list of outputs that appears in the statement of project scope is dominated by *ITO outputs*—those that emerge from application of the ITO model as outputs that must be utilised by project customers if target outcomes are to be generated. The panel of preferred suppliers in Project SmartDeal is a case in point.

In general, projects will also include other outputs which play little, if any part in utilisation—called *non-ITO outputs*—which can be usefully classified as

- **Mandatory outputs**: required by law, policy or regulation. Consider a scenario which might have emerged in the Project SmartDeal case, whereby the government passed a new law requiring that companies now report all orders on overseas suppliers exceeding \$1M. The scope of project SmartDeal would then have to be expanded with other outputs such as "annual import reporting process" and "imports tracking system".
- **Project set up outputs**: required to enable/support the work of project execution. Before execution can begin it is usually necessary to create some project management infrastructure—such as office space, computer systems and (on certain large exercises) worker accommodation. One notable example appeared during the construction of the Hong Kong airport at Chek Lap Kok in the late 1990/early 2000s. The sheer number of vehicles on site at any one time required the development of an entire site-specific transport management system including vehicle registration processes, tracking systems, a framework of regulations and a team of compliance officers. Such infrastructure will be peculiar to each project—in some cases involving a small number of modest outputs, but in others a large number of major outputs. The Chek Lap Kok site vehicle management infrastructure would appear in the statement of scope for the airport as a *project set up output*.
- **Stakeholder engagement outputs**: required so that particular stakeholders can be engaged in the project. Assume that the purchasing staff PKW Inc. are not entirely comfortable with Project SmartDeal, and that, to engage them, it has been decided to establish a performance bonus scheme. Such a scheme is an example of a *stakeholder engagement output*.
- **Risk mitigation outputs**: required so that a threat can be managed. Consider a project to build a light rail between a major transport hub and a popular recreational area in an area that is prone to occasional inundation from a local creek. To reduce

Table 9.1 Target outcome definition

Attribute	Meaning	Example
Name of target outcome	The measurable effect being sought by the funder	*Reduced order fulfilment error rate*
Description	Descriptive detail about the target outcome	*A fulfilment error occurs whenever there is a mismatch between an order, the shipping note and the invoice*
Measured variable	The name of the variable to be measured, e.g.: count, %, hrs, km	*% (monthly)*
Calculation	How is value of measurement derived?	*Error rate = (count of fulfilment errors/total shipments) * 100*
Source of data/method of measurement	Identify names of all: • Existing sources • (New) methods by which data will be acquired	• *Existing outbound logistics records* • *(A new) monthly order summary*
Target value	What is the target threshold being sought for this measure?	*A reduction of 25%*
Date for achievement	Expected achievement date for this outcome	*End November 2020*

the frequency of disruptions to construction work and, later, operational schedules, it has been decided to construct a system of floodgates across the creek. The flood control system would appear in the statement of scope for the rail project as a *risk mitigation output*.

It is sometimes difficult to decide if some outputs are utilised *directly* by customers. Take, for example, the IT infrastructure which must be assembled to run a website. While the website itself is utilised directly by project customers, the situation with the infrastructure is less clear. Any ambiguity about the status of the infrastructure is easily resolved by notionally combining it with the website into one output. In this case, the resulting "aggregated" output could be labelled as something like "website and supporting infrastructure".

9.3.4 Defining Target Outcomes

A target outcome is defined in a table with seven attributes—as shown in Table 9.1. Multiple target outcomes may appear as separate columns in a single table.

Some observations on this table

- **Name of target outcome**: As discussed in Chap. 2, this can be expressed in either *relative* or *absolute* terms. In relative terms, the outcome takes the form of a change in a variable *that currently exists* (The reader is reminded that, according to the ITO model, change is *scenario*-based, not *time*-based).
- **Description**: The description provides a supporting overview for the target outcomes. For example:
  - Is it confined to a particular geographic area or timeframe?
  - What is the meaning of any technical terms being used?
- **Measured variable**: Here is shown (only) the name of the variable being measured. Frequently this will be expressed as: "$", "count" or "%". If the variable being measured is attached to a *judgment* by someone, the measured variable appears as "Index"—i.e. it has no units of measurement. This would be the case if, for example, Likert scales are being used to gauge the target outcome "Increased employee satisfaction".
- **Calculation**: If the measured variable involves some sort of formula, then the formula is shown here. If the measured variable does not involve any mathematical manipulation, then this entry is left blank.
- **Source of data/method of measurement**: Here the origin of the data used in the measurement of the target outcome is identified. The data will either come from an existing source—or it will be specially acquired for the project. If the measured variable is calculated by a formula, then all sources of all variables used in the formula are identified.
- **Target value**: This is always shown as a simple number (or as a change in a number). No other detail is to be provided. This then becomes the threshold of the target outcome to which a commitment is being made in the business case.
- **Date for achievement**: This commits to a date by which the target will be achieved. This date will be later confirmed or revised in light of the project plan.

Note that once adopted, target outcomes are not required to be ranked, classified by importance or prioritised. They are either adopted for achievement or they are not. Some funders may, however, decide to emphasise some target outcomes as more important for them than others, to ensure full attention is paid by the project owner during the project for their realisation if trade-offs need to be made.

9.3.5 The Project Scoping Toolset

At any point in the process of refining of target outcomes and committed outputs, the two lists can be *validated*. In any event, the scope of the project must be validated before it can be used in the analysis. Here we examine the two tools used to validate a project's scope.

A. The outcomes coverage matrix (OCM) shows the relationships between the list of target outcomes and the list of downstream processes. There is *only one* OCM for a project. The OCM exposes any under- or over-scoping in a tentative scoping statement arising from an incorrect list of downstream processes.
B. Integrated utilisation maps (IUMs) show the relationships between the list of committed outputs and the activities in a downstream process. The activities appearing in the IUM for a downstream process define that downstream process.
   - There is one IUM *for each of the downstream* processes identified in the OCM.
   - Each IUM generates some of the FPFs which (eventually) define an output.
   - Collectively, the IUMs generate the FPFs which define each of a project's outputs and, as a byproduct of that, also expose any under- or overscoping in a tentative scoping statement due to an incorrect list of outputs.

9.3.5.1 The Outcomes Coverage Matrix

The Outcomes Coverage Matrix (OCM) confirms that a plausible causal relationship exists between the project and its target outcomes. Each row of the OCM is associated with one downstream process, while each column is associated with one target outcome. Entries are either 1 or 0—indicating whether the target outcome represented by the column serves as a performance metric for the downstream process represented by the row. This structure is based on the *principle of duality* (introduced in Chap. 2) which states that the utilisation of a project's outputs by project customers is equivalent to the execution of downstream processes by process agents (We use the term "process agent" here to mean those who participate in the execution of the downstream process). This observation confirms that the target outcomes from a project can *always* be expressed in terms of certain performance metrics observed during the execution of downstream processes.

We can infer from the principle of duality that if a target outcome does not represent an operational performance metric for at least one of the project's downstream processes, then it cannot be generated by the project (and hence the project is underscoped). Similarly, if none of the project's target outcomes serves as a metric for a particular downstream process, then that process is superfluous for the project definition (and hence the project is overscoped). Table 9.2 offers an example based on the Project SmartDeal case study for which there are two downstream processes: internal procurement and external procurement. Internal procurement is the process by which an operational or functional unit of PKW Inc. places an internal order for a product or service with the national procurement team. External procurement is the process by which the national procurement team places and manages an external order for a product or service with a supplier

An entry of "1" implies that the target outcome (at the top) is a performance metric of the process at the left. An entry of "0" means that the target outcome (at the top) is not a performance measure of the process at the left. In the discussion that follows we use matrix notation, whereby (i, j) refers to the ith row and the jth column.

Table 9.2 The outcomes coverage matrix (OCM) for project SmartDeal

Downstream process	Target outcome	
	Reduced payment times to suppliers	Reduced internal procurement time
Internal procurement	0	1
External procurement	1	0

For example, the "1" in cell (1, 2) (row one, column two) indicates that internal procurement time is a performance metric of the internal procurement process (as one would expect). By reengineering the internal procurement process, the company seeks to reduce the time taken to execute the internal procurement process.

By way of contrast, consider the "0" entry in cell (2, 2). The internal procurement time (the elapsed time between the submission of an internal order by a functional/operational unit and placement of a purchase order on a supplier) is not a metric of the external procurement process, because the internal procurement process must be completed before the external procurement process begins.

Consider the various patterns of "0" and "1" that might arise in an OCM for a (tentatively) scoped project

- **A column has only "0" entries**. This means that the associated target outcome is not a metric for any of the project's downstream processes and hence the project is incapable of generating that outcome. This implies, in turn, that the project is underscoped. On the (reasonable) assumption that the funder is committed to all of the outcomes in scope, then an additional downstream process will have to be identified which does have the missing target outcome as a performance metric. Such an addition will, in general, then require the expansion of current scope to include more outputs.
- **A row has only "0" entries**. This means that none of the target outcomes is a metric for the associated downstream process. In other words, the particular process is incapable of generating any of the project's target outcomes—which implies, in turn, that the downstream process is redundant and can be deleted from the OCM.

9.3.5.2 Integrated Utilisation Maps

An Integrated Utilisation Map (IUM)—shows when each committed output is utilised during execution of a particular downstream process, and what characteristics that output must have if it is to be successfully utilised.

All operational processes have a script—which in the case of a downstream process, takes the form of a series of relatively high-level sequential *activities*. The sequence of high-level activities appearing in an IUM *defines the downstream process* related to that IUM. Each *row* of the IUM is associated with one of the sequential activities that make up the script for the downstream process. In other words, the rows in the IUM are ordered in the same way that the activities are addressed during the execution of this particular downstream process. By way of contrast, the structure of columns is exactly the same for every IUM, whereby they are arranged (left to right) in the following way:

- **Seq #**. Indicating the order in which the activities are executed.
- **Activity description**.
- **Process agents**. Entries in these cells take the form of a *list* indicating all those project customers who are involved in execution of the activity.
- **A collection of columns entitled "outputs"**. Each of these associated with one of the project's committed outputs. While the ordering of these "output" columns is arbitrary, but it must be the same for all IUMs.

An entry in a cell of the IUM takes the form of *a list of fitness-for-purpose features* (FPFs) for the output identified at the top of the associated column. These FPFs show all the characteristics of the output that are critical to execution of the activity identified at the left of the associate row. There are three sorts of cell entry in an IUM:

- **The cell simply displays "N/A" (not applicable)**: This means that the associated committed output *plays no part* in the execution of the associated activity.
- **The cell simply displays the word "Generic"**: This means that the output *is utilised* during execution of the activity—however, beyond the generic FPFs normally expected in such an output, it need have no other special characteristics.
- **The cell has a list**: this indicates that the output *is utilised* during execution of the activity—and that, the listed FPFs are required *in addition to* any generic characteristics normally expected in such an output.

As suggested in Table 9.3, the structure of an IUM also allows for:

- Some additional rows at the top (to identify the project and the downstream process).
- A column to identify the entities involved in execution of each activity. These participating entities are called "process agents".

Table 9.3 shows a "fragment" of an IUM for the downstream process called "Internal procurement" drawn from the Project SmartDeal case study. For simplicity of discussion, this is a reduced version of the much more complete version of the IUM that would be expected in a real-life business case, in which

- The downstream process "Internal procurement" is represented by a small number of (unrealistically) high-level activities.
- Only three of the Project's SmartDeal's committed outputs appear.

The reader will note that an FPF for an output may be repeated. FPFs for a specific output may be duplicated in two ways: in different cells for that output within an IUM or across different IUMs. This means that, in general, a concatenated list of all the raw FPFs generated for an output by all of a project's IUMs will include a large number of redundant items. A raw list of that kind must be rationalised so that each FPF occurs only once. The resulting rationalised list then *defines* that output. The

Table 9.3 (A "fragment" of) one of the IUM's for project SmartDeal

Name of project:		Project SmartDeal			
Name of downstream process:		Internal procurement			
Seq #	Activity description	Process agent(s)	Outputs		
			New policy and procedures manual	Restructured procurement unit	Panel of preferred suppliers
1	Register need for products or services	Local functional/operational unit	Version for local management	New unit to have teams devoted to particular types of supply (e.g. trucks, stationery)	N/A
2	Acknowledge internal order	The procurement unit	Version for procurement staff	Generic	N/A
3	Issue periodic order status report	The procurement unit	Generic	Generic	N/A
4	Confirm receipt of purchased items services	Local functional/operational unit	Generic	Generic	N/A
5	Close internal order	Local functional/operational unit; The Procurement unit	Generic	Generic	N/A

following overview of the outputs definition summary explains the rationalisation process.

9.3.6 *Defining Committed Outputs*

A committed output is defined by a list of FPFs. For ITO outputs, these lists are drawn from the project's IUMs. In the case of non ITO outputs, they are assembled with reference to their intended purpose. The FPFs for mandatory outputs will be drawn from the law, policy or regulations which dictates production of the output.

The FPFs for stakeholder engagement and risk management are drawn intuitively from the respective register in which they have been identified.

The process for defining a particular ITO output is based on these steps (done separately for each output)

1. From each IUM extract all the entries related to the output in question.
2. Concatenate all these entries into a single list—which may have duplicated entries.
3. Eliminate all duplicated FPFs (if they exist).
4. Eliminate all the entries appearing as either "N/A" or "generic"

The resulting rationalised list of FPS defines the output.

If the list of FPFs is "wrong" then the output will have been defined incorrectly. There are two ways in which an output can be defined incorrectly

- **"Under-engineered"**, whereby the definition is missing one or more *critical* FPFs. A critical FPF is one that is necessary for successful utilisation of the output by a project customer.
- **"Over-engineered"**, whereby the definition includes not only *all* critical FPFs, but also other (superfluous) FPFs.

(Note that an output definition that is missing a critical FPF is classified as under-engineered, even if it happens to include superfluous FPFs).

If an output is under-engineered, then utilisation will be impeded. At the very least this will hinder achievement of target outcomes. If an output is over-engineered, then the superfluous FPFs may (pointlessly) increase the costs of producing it.

In the template for a *project* definition (See Box 9.3), Sect. 9.3.2 shows, for each output, not only a list of *included* FPFs (which define that output), but also a list of *excluded* FPFs. The rationale for this second list is similar to that for the items which appear in Section 6 in Box 9.3—Scope boundaries (Sect. 9.3.2). If early discussions with those who can help shape the project generate expectations that particular outputs will have particular characteristics, it is critical that such expectations be properly managed. The following example shows one of the outputs from the Project SmartDeal case study with some illustrative included/excluded FPFs.

Panel of preferred suppliers

Included FPFs

- Members classified by range of goods/services
- Each member equipped with appropriate systems and infrastructure

Excluded FPFs

- Members to be certified under ISO 9000

Note that when shown in this way, the definition of a committed output takes the form of a three-level hierarchy in which

1. Level 1 is a list of all the projects committed outputs.
2. Level two has only two values: "included FPFs" or "excluded FPFs.
3. Level 3 is a list of the corresponding FPFs.

9.3.7 Setting Boundaries for Project Scope

Early in initiation, the project champion (together with the person commissioned to assemble the document) will engage with a wide range of potential stakeholder-s—looking for candidate target outcomes, outputs and downstream processes. This usually results in a draft scoping statement with very long lists—many of which will not be adopted for inclusion in the project as eventually proposed. Eventually, these lists must be culled—so that they typically finish up with no more than five target outcomes—but with as many outputs and downstream processes as are necessary to generate the target outcomes. All those who have been involved in initiation will have discussed (or perhaps even suggested) various candidate outputs, outcomes and downstream processes, many of which will not be included in the eventual project proposal. To ensure that expectations are managed properly, and that there is no misunderstanding, all items that have been discussed—but then discarded, should be specifically identified in the relevant three lists

A. Excluded target outcomes (names only—no details).
B. Excluded outputs (names only—no details).
C. Excluded downstream processes (names only—no details).

(In each case, all excluded items must be labelled in accordance with the relevant word structure).

9.3.8 Target Outcome Baselining

In Sect. 2.4.4, the discussion of the 2NY diagram raised the issue of baselining. When relative naming is being used for target outcomes ("Increased ..."/"decreased ..."), how is a target to be set for the implied change? Take for example a project by an online business to reduce order fulfilment time. In the project definition, a target value must be set, but at what value? To resolve this issue, the funderFunder, first of all, needs to know how long it currently takes to fill an order (It might already be so low, that a further reduction is impractical). Baselining is the process of measuring the "Now" values for target outcome variables.

Ideally baseline values for target outcomes will already be known from existing data sources. The online business may, for example, already record this during execution of the current order fulfilment process. If such data is not already available, then there are three alternative strategies for its acquisition—each of which bears further discussion

A. Undertake a baselining exercise as part of project initiation
B. Conduct baselining in parallel with planning
C. Conduct baselining in parallel with execution

Regardless of the option, baselining involves additional costs. Strategy A is the most desirable because it means that the in-principle funding decision at the end of initiation will be based on complete information about the project's worth. So why might it make sense to delay baselining until after initiation is complete (options B and C)? Furthermore, what are the implications of such a decision? A delay may be merited if two conditions are met

- The funder is confident that a "significant": change is achievable
- An "early" qualified decision to proceed is seen by the funder preferable to a "late" unqualified decision to proceed

In effect, under options B and C, the funder is taking a gamble that baselining will eventually confirm the appropriateness of the decision to proceed.

9.4 Project Analysis

Analysis is the third sub-process in the project initiation phase.

9.4.1 Overview of Analysis

During this sub-process, the project definition is analysed to obtain reliable values for various parameters used to determine a project's worth. In the course of this analysis, additional contextual information is uncovered (or simply becomes available) about such project elements as stakeholders, risks and issues. All this, together, with the definition itself, allows the funder to gauge the project's attractiveness for investment. In this section, while we are primarily concerned with the techniques used to determine: timeframe and costs, we also consider other items of contextual information required for the business case.

While the same techniques for estimation of time and cost are employed in both initiation and planning, they are used to significantly different levels of intensity (It is for this reason that we expect the uncertainty in estimates produced during planning to be less than the corresponding figures produced during initiation). Because planning is dominated by the work of estimating time and cost, we cover the associated tools and techniques more thoroughly in Chap. 10. Their treatment below is somewhat cursory—going into only as much detail as required to make sense of project analysis. We leave it to the reader to delve into Chap. 10 as the need arises.

In what follows we cover, in turn

- Estimating the duration of project execution
- Estimating the cost of project execution
- Budgeting and cashflow planning
- Updating the (three) project registers (see Sect. 9.4.5).
- Assembling a project communications strategy
- Updating the project governance model

The focus of analysis is on (eventual) project execution; however, the timeframe and cost of *the project overall* are also determined by the timeframe and cost of *planning*—the global phase which separates initiation from execution. So how are those parameters taken into account? The reader will note that the last section of the business case is entitled: "An overview of planning". As Sect. 9.4.2 discusses, it is here that estimates of the timeframe and cost of planning are provided (as separate figures) for the funder because they represent the time and cost involved in making a final commitment to an investment in the project (or abandoning it).

Estimates of project-level timeframe and cost are derived by considering both those parameters for each output on its own. The cost of the project can then be obtained by simply summing the costs of all outputs and combining those with estimates of the above-the-line resources anticipated during execution. However, because outputs can be produced in parallel, the timeframe for the overall project will, in general, be significantly less than the sum of the timeframes of all committed outputs. There are two broad approaches to estimation of both timeframe and cost, top-down and bottom-up. Top-down estimation is attractive because it looks at the actual experience with similar outputs that have been produced in past projects and adjusts them for differences in the current exercise. In practice, this is the most common approach to estimation. When it is not appropriate to use that technique, bottom-up estimation may be employed. This involves assembling a detailed "model" of the script for the work involved by breaking it into successively smaller "chunks". The time and cost of completing each *chunk* can then be aggregated into corresponding estimates for the *output* under consideration.

The following discussion is about *bottom-up* estimation. Chapter 10 not only expands this into considerable detail, but it also explores top-down estimation in more depth.

9.4.2 Estimating the Duration of Project Execution

To obtain a bottom-up estimate of the overall timeframe for execution, the duration of the work required for each output is estimated and then interdependencies used to expose: those tasks that can be conducted in parallel and those that cannot. All this results in a workplan which has three elements

- A work breakdown structure (WBS): which is a *hierarchical model* of the work on each output).
- A Gantt chart (a diagram in which each component of the WBS is represented by a bar which is overlaid on a horizontal calendar).
- A schedule of milestones (SoM): in which critical dates drawn from the Gantt are assembled into a schedule for later use in tracking the progress of execution.

In practice, the WBS does not appear separately but is incorporated into the Gantt chart, while the SoM is usually shown as a separate table. For many project outputs, a stylised three-level "PAT" (phase-activity-task) hierarchy is adequate for a WBS during initiation (a more detailed WBS that requires more than three levels of hierarchy may be required during project planning)

- At level 1, the work of producing the output is broken into *phases*.
- At level 2, each phase is broken into *activities*.
- At level 3, each activity is broken into *tasks*.

The reader might note that the *three-level* WBSs of all committed *outputs* create a *four-level* WBS for the *project*: output, phase, activity and task.

Because a WBS is a model of *work*, rather than a model of an *output*, we impose a work structure on the labels attached to tasks—whereby they take the form of an *imperative* (a command or an instruction). Some examples of correctly named tasks in a WBS

- *Assemble* model of sales process.
- *Weld* bracket to column.
- *Conduct* concrete compression test.
- *Assess* performance of prototype.

A stylised PAT-style WBS (made up of phases, activities and tasks) can accommodate about 350 tasks—which is suitable for many typical outputs from a project. The rationale for "approximately seven" (which, for simplicity, we call "The rule of 7-ish"), is discussed in 10.2.2.

The WBS can now be displayed as a Gantt chart—in which each WBS element is represented by a horizontal bar laid over a calendar. To do that one must

- Estimate the duration of each task.
- Note any dependencies among the tasks
- Locate each task diagrammatically on a calendar—constrained by its dependencies

Figure 9.2 shows an incomplete "fragment" of the WBS and Gantt chart associated with the landscaping of a domestic garden.

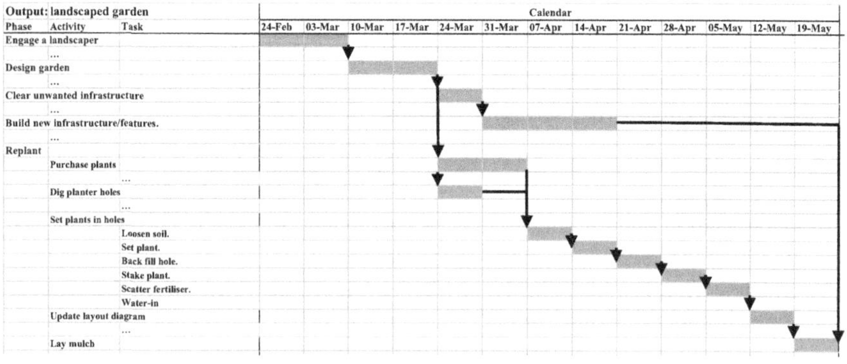

Fig. 9.2 An illustrative portion of the Gantt chart for Project SmartDeal

Table 9.4 A schedule of milestones drawn from the Gantt chart shown in Fig. 9.2

#	Milestone name	Date
A	Engage a landscaper	10-Mar
B	Design garden	24-Mar
C	Build new infrastructure/features	21-Apr
D	Hire backhoe	21-Apr
E	Excavate holes	28-Apr
F	Set plants in holes	12-May
G	Mark plant locations	19-May

The arrows indicate precedence. Notice how the tasks in the Gantt chart form a number of "chains" of connected tasks. It is the longest of these chains which determines the overall duration of the work to produce the output (a landscaped garden). In this case, the longest chain runs over 13 weeks:

- Design garden
- Engage landscaper
- Build new infrastructure/features.
- Hire backhoe
- Excavate holes
- Set plants in holes
- Mark plant locations

This "longest" chain is called the *critical path* because delays in any of the tasks which form the chain will delay delivery of the garden by the same amount. For that reason, later (during execution), a selection of the tasks on the critical path (which need not be all of them) will be monitored closely to confirm that each has been completed by the due date.

The selection of tasks from the critical path used for this monitoring is assembled into a table called the *schedule of milestones*.

Table 9.4 shows a schedule of milestones drawn from the Gantt chart in Fig. 9.2. Note that a milestone is *a point in time*—defined by *the completion date* of the associated task on the critical path.

9.4.3 Estimating the Cost of Project Execution

Recall that project costs arise from the resources consumed by the work of the project. While the most obvious costs are those represented by cash outlays for external resources, internal resources (such as seconded staff) must also be recognised as a cost.

To obtain a bottom-up estimate of the overall cost execution, the costs attached to each output are estimated and summed. The resources consumed during a project are

Table 9.5 An illustration of an internal resource plan (in person-days)

Resource name	Feb	Mar	Apr	Total
Home owner	2	6	1	9
Family members	5	5	3	13
Total	7	11	4	22

Table 9.6 An illustration of entries in an external resource plan ($)

Item name	Feb	Mar	Apr	Total
Landscaper	2,000	3,500	5,000	10,500
Backhoe	1,000	1,000	1,000	3,000
Watering system	0	2,500	1,000	3,500
Plants	500	750	1,000	2,250
Mulch/stakes	0	850	250	1,100
Total	3,500	8,600	8,250	20,350

usefully split into two: internal and external. Internal costs arise from any existing resources that will be allocated (deployed) to the project. The most common form of the internal resource is represented by staff from the funding organisation who are to work on the project. External costs arise from those resources that must be purchased for the project.

Because staff get paid whether they are appointed to the project team or not, in general, internal resources do not give rise to direct cash outlays for the project (This issue is covered in more detail in Chap. 10). Nevertheless, the demands on such staff must be estimated because, at the very least, while working in the project they are unable to carry out their functional/operational roles. To manage this, the organisation needs to know, for all staff: what demands will be made on their time, when they will happen, the patterns of those demands and their levels of intensity.

The simplest way of displaying estimates for internal labour is a table—along the lines of Table 9.5.

The bottom-up approach to the estimation of cost considers each task in the WBS and catalogues the items that must be purchased to complete that work. As was the case for internal labour, a table is the simplest way of summarising this information along the lines of Table 9.6.

9.4.4 The Project Budget and Cashflow Planning

At some point after approval of the project plan, preparations must be made to cover all of the outlays to be made during project execution. The business case includes a preliminary version of what will appear in the project plan. This allows the funder to consider the financial implications of undertaking the project. On its own, a single all-up estimate of project costs is of very limited value because that budget masks the pattern of period-by-period outlays during execution. Consider two similar revenue-generating projects "A": and "B", each with a budget of $1 M. "A" requires that the bulk of outlays be made early in execution. It also generates no revenue until execution is complete. "B", on the other and, requires that the bulk of outlays be made late in execution. It also generates some revenue during execution. The funder must have access to the full $1 M before work gets underway, while the funder of B may never need a pool of more than, say, $500,000.

A distinction should be drawn between the project's *cost* and its *budget*. A figure for all-up cost is required to determine the project's worth. This is based on the value of all (cash) outlays on purchases together with an imputed value for all internal resources. The cashflow table summarises the external resource plan in monetary units and so can be used directly when setting the project budget.

An effective way of presenting this sort of detail to the funder is to provide a cashflow table such as illustrated in Table 9.7 in which

A. Columns are associated with time periods.
B. Rows appear in two "blocks":
   - Cash inflows (broken out by type of flow if necessary)
   - Cash outflows (broken out by acquired resource).
C. Entries indicate the money involved in each case.
D. Column totals show the net cashflow for the period.
E. Row totals show the total inflows and outlays on each resource item

The grand total of all cash outflows is the project's budget.

The outlay rows would normally be a more aggregated version of those appearing in the internal and external resource plans.

Table 9.7 A cashflow illustration (values in $)

Item	Period 1	Period 2	...	Period N	Totals
Inflows					
Government research grants	0	100,000		0	100,000
Revenue	0	50,000		250,000	300,000
Inflows sub totals	0	150,000		250,000	400,000
Outflows					
Land acquisition	750,000	0		0	750,000
Construction	0	125,000		20,000	145,000
Fit out	0	15,000		200,000	215,000
Outflows sub totals	750,000	140,000		220,000	1,110,000
Cashflow	−750,000	10,000		30,000	−710,000

9.4.5 Assembling and Maintaining the Registers During Initiation

Although they play no part in the definition of a project, the three registers should be available throughout that sub-process. They can then be used to catalogue candidate entries as they are identified. Other than the name of the potential stakeholder, risk or issue, no other detail need be provided at this time (In other words, only the first two columns of each register will contain data).

These partly populated registers then provide a useful starting point for the assembly and maintenance of the registers—work that must be done progressively during the analysis sub-process. This will ensure that the funder has reliable versions of the registers when deciding whether to accept the business case.

9.4.6 Assembling a Project Communications Strategy

Because they are handled differently, it is useful to distinguish between internal and external project communications. Internal communications are those between the project and everyone who fills a role in the project governance model. External communications are directed at who has a spontaneous stakeholding in the project (regardless of whether they fill a role in the project).

With one exception, the internal communications framework for a project is stylised and somewhat formulaic. Chapter 11 covers this topic in more detail. The exception relates to the especial demands for communication noted in a role definition. The generic role definition template shown as Table 4.1 includes an item "Communications programme". This identifies any special forms of communication (beyond those embedded in regular project reporting) that need to be instituted for the role in question. These are incorporated into the same communications planning framework proposed below for spontaneous stakeholders.

The stakeholder register includes three possible classes of engagement—the first of which is "communicate with the stakeholder". For every entry in the stakeholder register, there will be **at least one** recommended form of communication. If we extracted all such entries from the stakeholder register, *as well as* from the "communications programme" entries in role definitions, they could be summarised as a two-level hierarchy along the lines of Table 9.8.

Because each entry in the stakeholder register is handled independently of the others, the sort of repetition displayed in this table is common. A communications strategy is assembled by inverting the hierarchy so that instead *of showing forms of communication under each stakeholder*, it shows *stakeholders under each form of communication*. This can now be elaborated with details about what features must be built into each form of communication to meet the peculiar needs of each stakeholder. The result is a communications strategy—like that in Table 9.9.

Table 9.8 An illustrative communications extract from a stakeholder register

Stakeholder	Form of communication
Fred Nurke	Provide with access to project website
Jane Dickinson	Email periodically to seek feedback
	Invite to monthly project briefing
	Provide with access to project website
Members of Club Waziema	Provide with access to project website
	Send monthly project newsletter
Neddy Seagoon	Invite to monthly project briefing
	Send monthly project newsletter
	Provide with access to project website
Elaine Spencer	Provide with access to project website
	Email periodically to seek feedback

Table 9.9 An illustrative communications strategy based on Table 9.8

Form of communication	Stakeholder	Necessary features
Project website	Fred Nurke	Provide weekly status reports, show updated completion date
	Jane Dickinson	Highlight work being undertaken to reduce noise from site, highlight current/imminent street closures
	Members of club Waziema	Offer choice of English or Amharic
	Neddy Seagoon	Use graphics instead of text where possible, use simple words and short sentences
	Elaine Spencer	No special features
Periodic feedback email	Jane Dickinson	Gauge level of support for project
	Members of Club Waziema	To be in both English and Amharic
Monthly project briefing	Jane Dickinson	Cannot be held on Mondays or Tuesdays
	Neddy Seagoon	Follow with casual gathering over dinner, hold at a local hotel
Monthly project newsletter	Members of Club Waziema	To be in both English and Amharic
	Neddy Seagoon	No more than two A4 pages

The communications strategy provides no details about the design, development of delivery of each form of communication—that will be done early in execution (based on any necessary updating carried out during planning).

9.4.7 Assembling a Project Governance Model

Together, the funder, the project champion and the person assembling the business case will consider the type of organisational structure most appropriate for the project's eventual execution, and then progressively assemble a project governance model suited to the peculiarities of the project. In doing this, first of all construct a PGM diagram which will show all the roles to be recognised in execution and then, for each position, create a role definition. In some cases, the names of the entities filling these roles will be known—in which case that information is displayed in the PGM diagram and recorded in the role definition. Most role definitions will be based on a template (along the lines of Table 4.1) in which all the details appearing in the "Description" will be standardised for the funding organisation. In other words, each project will draw in a library of generic role definitions covering: project owners, project managers, steering committee members and so on. The most important issues to be considered here are

- Who will serve on the steering committee?
- Who are to be appointed (if not already) as project owner and project manager?
- Is the project big enough to justify a project control group?
- What reference groups/advisers are to be established?

A further note about the last of these questions is appropriate. Recall from Chap. 4 that there are two types of reference group/adviser

A. Those created to establish a positive relationship between a spontaneous stakeholder and the project's key players. Here the engagement is purely a form of stakeholder engagement.
B. Those created to gain access to them as a project resource, regardless of whether or not that entity is a spontaneous stakeholder in the project.

In both cases, a customised role definition must be written for each reference group and adviser. For case "A" above, this will draw on relevant entries appearing in the stakeholder register under the heading "Engagement programme"/"Have the stakeholder participate in the conduct of the project".

9.5 Assembling the Business Case

9.5.1 Overview

When the sub-process of analysis is complete, all of the components of the business case have been created. Now only the packaging sub-process remains—which then leads to the tabling of the document by the champion and its appraisal by the funder.

Packaging involves gathering together all these components, ensuring that they are complete and rationalising them so that they are all internally consistent. In most cases, the tabling of the document would be followed by some form of presentation.

9.5.2 Packaging the Business Case

Before they are integrated into one document, some of the components may need to be tidied up. Checking for consistency with other components is an important part of this tidying up.

1. **Introduction**: may not have yet been written. Completing it requires no additional analysis.
2. **Business context**: may not have yet been written. Completing it requires some relatively straightforward analysis.
3. **Project definition**: will be complete.
4. **Project Governance**: some additional updating may be demanded by recent developments (such as additional appointments to fill various roles).
5. **Managing the External Environment**: the dynamic natures of stakeholding, risk and issues mean that continuous updates of the registers will be required—right up to the point where the business case is tabled.
6. **Project analysis**: if done thoroughly, this is unlikely to require any specific updating.
7. **An overview of planning**: may not have yet been written. Completing it requires no additional analysis.

At this point, the document would become the subject of the organisation's standard procedures and business practices

9.5.3 A Template for a Business Case

Box 9.4 shows the structure of the business case—in the form of a template. This structure is explained in the following section.

Template

Box 9.4: The Business case

1. **Introduction**
   - 1.1 Name of project
   - 1.2 Purpose of business case
   - 1.3 Overview of project
   - 1.4 The project's key players
2. **Business Context**
   - 2.1 The thrust of the project
   - 2.2 Rationale
   - 2.3 Organisational impact statement
   - 2.4 Analysis of options
   - 2.5 Related projects and programmes
   - 2.6 Assumptions and constraints
3. **Project Definition**
   - 3.1 Statement of scope
   - 3.2 Target outcome definitions
   - 3.3 Validation of project scope
   - 3.4 Committed outputs definitions
   - 3.5 Scope boundaries
4. **Project Governance**
5. **Managing the External Environment**
   - 5.1 Stakeholders
   - 5.2 Risk
   - 5.3 Issues
6. **Project Analysis**
   - 6.1 WBS/Gantt chart
   - 6.2 Schedule of milestones
   - 6.3 Internal resource plan
   - 6.4 External resource acquisition plan
   - 6.5 Project budget
7. **An overview of planning**

9.5.4 The Structure of the Business Case

Here we discuss each section of the template provided in Box 9.4.

1. **Introduction**: provides a setting for the business case

   1.1 **Name of project**: A project's name should clearly identify it and distinguish it from others in the funder's portfolio of projects.

   1.2 **Purpose of business case**: In general, this will be worded similarly for every project.
   "The purpose of this business case is to:
   a. *Provide all the information required by < the funder >to make a reliable in-principle funding decision.*
   b. *Establish the project's key parameters.*
   c. *Brief key stakeholders about the project"*.

   1.3 **Overview of project**
   A. A concise overview of the project so that the reader gains a quick understanding of its overall "shape". The description should be succinct because the following sections provide all the necessary detail.
   B. A short appraisal—summarising all of the parameters that determine the project's overall attractiveness: benefits, disbenefits, costs and riskiness (by giving them each a quantitative or qualitative value—such as *high*, *medium*, *low*). This appraisal could be further broken into a sequence of brief discussions about:
      - Return (based on worth and as a function of benefits, disbenefits and costs).
      - Riskiness.
      - Attractiveness (as a function of return and riskiness)
   C. The wording of this section will depend on the audience, the extent of their background knowledge nature the initiative itself.

   1.4 **The project's key players**
   Here identify and briefly discuss the role of
   a. The funder(s)
   b. b.The champion (or the project owner if already appointed).
   c. The person who assembled the business case (under the guidance of the champion).

2. **Business context**:

   2.1 **The thrust of the project**: a high-level descriptive summary of the project's scope (which is formalised in Sect. 9.3.1)
   a. What will success look like?
   b. What are the most prominent outputs?

   2.2 **Rationale**: A brief summary of what is driving the project—covering such questions as:
   a. A brief history of events that have led to this business case.

b. Briefly, what sequence of events have led to this business case.
c. Why this particular project at this particular time?
d. How does the project relate to the organisation's declared strategies/policies?
e. How does the project rank amongst competing initiatives?

2.3 **Organisational impact statement**

a. **Resource implications**. What impacts will execution of this project have on the funding organisation (especially demands for internal resources), and how will those impacts be managed?
b. **Strategic implications**. Does the project have any long-term implications for the organisation (other than the benefits/identified below?). What are those implications and how should they be addressed?
c. **Undesirable outcomes**. *Many projects have undesirable outcomes (even if only inconsequential).* These should be drawn to the attention of the funder. They should all be listed, described and those who will be negatively impacted clearly identified.

2.4 **Analysis of options**

Were any alternatives to this project considered? If so, briefly outline them and summarise the process that led to the adoption of this project as the preferred option. (A statement of project scope for any such options should be attached as an appendix to the business case).

2.5 **Related projects and programmes**

To what other projects and programmes is this initiative related and how will those relationships be managed?

Three links are of interest—other projects or programmes (completed or current):

a. … on which this project depends.
b. … which are interdependent with this project.
c. … which are dependent on this project.

If the proposed project interacts with linked projects, then a decision must be taken to either coordinate them—or run them independently. If they are to be coordinated, then this is achieved by incorporating them into a programme (See 4.2.1 and 4.2.2).

2.6 **Assumptions and constraints**

What *values of key variables* have been assumed? *What situations/conditions* have been assumed? Is it conceivable that any of these assumptions might be proven incorrect? What are the implications?

What limitations/restrictions have been imposed on the conduct of the project? Is it conceivable that any of these constraints might be violated? What are the implications?

3 **Project definition**

3.1. **Statement of scope**

This has five elements—as discussed in Table 9.2 above

a. Project objective.

b. The funder.
c. A list of target outcomes.
d. A list of committed outputs.
e. A list of the project's downstream processes.

Each of the three lists here provides the names of the items (using appropriate word structures) *without any other detail* (which appears elsewhere in the business case).

3.2 **Target outcome definitions**

Each target outcome is defined using seven "attributes". Because there will rarely be more than five target outcomes, their definitions are best represented by a table in which they appear as columns, while the rows show the attributes—along the lines of Table 9.1. The table is preceded by a general descriptive summary.

3.3 **Validation of project scope**

The three lists appearing in the statement of scope must be validated (that is, confirmed as "correct"). The two tools described in *9.2.4.1 The Outcomes Coverage Matrix* ("OCM") and *9.2.4.2 Integrated Utilisation Maps* ("IUMs") are used to validate project scope. There is one OCM for the project and one IUM for each downstream process. These are introduced with a general overview and presented as tables in this section.

3.4 **Committed outputs definitions**

An output is defined by its FPFs—all appearing as an inset list beneath the name of the output to which they apply. The list of FPFs is abstracted from all the FPFs appearing in all the IUMs for each output.

a. Those which are *included* in the definition.
b. Those which are specifically *excluded*.

This section, therefore, is made up of a number of such lists—one for each committed output.

The list of FPFs for an output is divided into two parts

3.5 **Scope boundaries**

The boundaries of a project are set by highlighting those items which have been specifically excluded from the statement of scope. This section is also made up of three simple lists

a. A list of any candidate target outcomes which have been *excluded* from the scope.
b. A list of any candidate outputs which have been *excluded* from the scope.
c. A list of any candidate downstream processes which have been *excluded* from the scope.

4. **Project governance**

This section defines the organisation model to be adopted during execution. It has two parts

- The project governance model (as a diagram) supported with a brief description.

- A set of role definitions for all the roles identified in the PGM.

Because the flow of the business case would be interrupted if those roles definitions were inserted into this section, a hyperlink can be used instead, or they may be attached as an appendix.

5. **Managing the external environment**

This section summarises the way in which the impacts of events arising from outside the project will be managed. The details provided here are ultimately drawn from the three registers: stakeholders, risk and issues. It is impractical to incorporate them into the text of the business case for two reasons

- They usually take the form of (relatively large) spreadsheets.
- Not all the information they contain is of significance for a funding decision.

To deal with this, for each of stakeholders, risk and issues three levels of detail are provided in this section

5.1. **Stakeholders**

5.1.1 **Stakeholder management**

a. A descriptive summary of critical information derived from the stakeholder register.
b. A hyperlink to the stakeholder report.
c. A hyperlink to the stakeholder register.

5.1.2 **Project Communications strategy**

The project communications strategy takes the form of a table such as that shown above as Table 9.9. If that is so big that it interrupts the flow of the document, then it may be attached as an appendix (with an appropriate reference inserted here).

5.2 **Risk**

a. A descriptive summary of critical information derived from the risk register.
b. A hyperlink to the risk report.
c. A hyperlink to the risk register.

5.3 **Issues**

a. A descriptive summary of critical information derived from the issues register.
b. A hyperlink to the issues report.
c. A hyperlink to the issues register.

6. **Project analysis**

This section provides the results of a high-level analysis of the project definition

6.1. **WBS/Gantt chart**

In general, this will take the form of a workplan created with an appropriate scheduling tool (such as Microsoft Project).

This section will include

a. A brief descriptive summary of any noteworthy highlights.
b. A hyperlink to the relevant scheduling file.

6.2. **Schedule of milestones**
This will take the form of a table—such as that suggested in Table 9.4. In general, it will be small enough to show here in full.
This section will include
a. A brief descriptive summary of any noteworthy highlights.
b. The schedule of milestones itself.

6.3 **Internal resource plan**
This will take the form of a table—such as that suggested in Table 9.5. In general, it will be too big to show here in full.
This section will include:
a. A brief descriptive summary of any noteworthy highlights.
b. A hyperlink to the relevant internal resource demand table.

6.4 **External resource acquisition plan**
This will take the form of a table—such as that suggested in Table 9.6. In general, it will be too big to show here in full.
This section will include
a. A brief descriptive summary of any noteworthy highlights.
b. A hyperlink to the relevant external resource table.

6.5 **Project budget and cashflow planning**
This will take the form of a table—such as that suggested in Table 9.7. In general, it will be too big to show here in full.
This section will include:
a. A brief descriptive summary of any noteworthy highlights.
b. A hyperlink to the relevant cashflow table.

7. **An overview of the planning global phase**
If the funder decided to progress the project to the second global phase (planning), he/she needs to know about the timing and resourcing implications. This section will provide that information by summarising:

- How long planning will take.
- Who should be involved? Doing what? When? How much of their time will be required?

9.5.5 *The Impact of Planning on the Business Case*

As will become clearer in Chap. 10, the project plan *in toto* is a baseline document—in that everything it contains is used to guide execution. Here we consider the question: "How much of the original business case can be used to augment the project plan as a baseline?".

Table 9.10 reveals that the only part of the business case that is: (a) relevant to the project's baseline and (b) unaltered by planning is the *project definition*. All other parts of the business case which might otherwise have filled a baseline role are now

Table 9.10 The implications of planning for the business case

Business case section and title	Implications of planning
1. Introduction	
1.1 Name of project	Irrelevant to execution
1.2 Purpose of business case	Irrelevant to execution
1.3 Overview of project	Possibly updated
1.4 The project's key players	Possibly updated
2. Business context	
2.1 The thrust of the project	Remains unaltered
2.2 Rationale	Remains unaltered
2.3 Organisational impact statement	Completely reworked
2.4 Analysis of options	Remains unaltered
2.5 Related projects and programmes	Possibly updated
2.6 Assumptions and constraints	Possibly updated
3. Project definition	Remains unaltered
4. Project governance	Significantly expanded and updated
5. Managing the external environment	Significantly expanded and updated
6. Project analysis	Completely replaced by the project plan itself
7. An overview of planning	Irrelevant to execution

included in the structure of the project plan. Therefore, at the end of planning, the project's baseline is based on:

- The project definition.
- The project plan.

9.5.6 Appraising the Business Case

The funder now has to decide if this project appears worthy of investment and if so—approve progress to the planning global phase. Bear in mind that the project owner will have kept the funder fully briefed on initiation as it progresses. If this has been done professionally, both will already have a reasonably clear idea about the acceptance or rejection of the business case

In doing so the two major points to confirm are:

- Does the document make an appropriate trade-off between return and risk? In other words, is the project correctly located on the project attractiveness map?
- How does this project rank with projects that are competing for investment funds?

The objective of initiation is not to have the project accepted—but instead is to ensure that a robust decision can be made about the acceptance (or otherwise) of the project. If the appraisal of a business case enables the funder to make a confident decision not to proceed, then initiation has achieved its target outcome—just as much as if a confident decision is made to proceed.

Because planning may well result in changes to certain critical project parameters (especially those related to worth), when approving the business case, the funder should indicate a range for each. These ranges establish bounds on the owner's delegated authority. If the value of a project parameter revealed by planning (or at any stage of project execution) falls outside this range (either above or below), then the project owner would be required to have the new figure approved by the funder. Different organisations and funders will have dissimilar such bounds, depending on factors such as the trust versus control personal characteristics of the funder, the financial situation of the company, and the level of project risk.

Summary

Initiation seeks to establish a robust definition of a project—which can then be used to obtain and package all the information required by a funder to make a decision about whether to fund a proposed project and progress to the planning global phase. The next chapter explores the planning process and the structure of the project plan—on which subsequent execution will be based.

Chapter 10
Planning a Project

Abstract When a funder accepts the business case for a proposed project and a project manager is appointed, it then progresses from initiation to the second global phase—planning. It is here that a script for the work (to be later undertaken during execution) is created. Chapter 9 discussed initiation in terms of four subprocesses (identification, definition, analysis and packaging). Planning is similar to the analysis subprocess in initiation but carried out in considerably more detail. This chapter discusses the major elements of project planning, the most effective tools and essential decision-making by project leaders.

10.1 Overview of Planning

When a funder accepts the business case for a proposed project, a project manager is also appointed. The first responsibility of the project manager is to plan the project in detail (Globerson and Zwikael 2002). It is here that a script for the work to be undertaken during execution is created. The script has three parts: A Work Breakdown Structure (WBS), a Gantt chart showing when each task in the WBS will be undertaken and schedule of milestones (used during execution to monitor progress). The focus of planning is the same ground covered at a high level in Sect. 6 (project analysis) of the business case; however, because this is now done in considerably more detail, it effectively replaces much of the business case. The output from planning is the project plan, which, together with the project definition appearing in the business case, provides the baseline documentation for execution.

Planning involves a combination of creative and analytical activities. The latter are supported by a variety of accepted practices, procedures, standards and computer-based tools (such as scheduling software).

This chapter begins by setting a context for planning. It then examines the form in which all of the work required during planning is presented—and the method underpinning that analysis. It then goes on to consider techniques for determining the timeframe and cost of execution, and finally discussing the importance of reliability as a criterion for judging the quality of a project plan.

O. Zwikael and J. R. Smyrk, *Project Management*,
https://doi.org/10.1007/978-3-030-03174-9_10

10.1.1 Planning in Outline

The prime focus of planning is on development of a script of project execution. This script includes:

- A catalogue of the tasks that must be carried out to produce the project's committed outputs.
- All dependencies amongst those tasks.
- Dates when each task is to be carried out.

From this script, we can then:

- Estimate the duration of execution.
- Identify the resources required to undertake project execution.

The word "estimate" means "a prediction" rather than a guess at a hidden value and so the central issue with planning revolves around the question "How are reliable estimates of timeframe and cost to be developed?". To ensure high reliability of the planning process, planning takes the committed outputs and all endogenous downstream processes (which also qualify as project outputs) and analyses each in the following sequence of steps which are undertaken iteratively (and hence largely in parallel):

1. A script is assembled—in the form of a Work Breakdown Structure (WBS) with a supporting Gantt chart. This script is, effectively, a model of the work required to produce each of the project's committed outputs. Amongst other things, this model allows the duration of execution to be estimated.
2. Partly informed by the Gantt chart, *top-down* techniques are most commonly used to estimate internal and external resources and develop an initial schedule for them (in the form of tables showing what resources are required and when). From this, the overall cost of execution can be estimated, and a cash flow developed for the project.
3. Certain details from both the Gantt chart and the resource schedules can then be combined into a schedule of milestones—on which the progress of execution will eventually be judged.

The work involved in outcomes realisation—the fourth global phase of a project involves a combination of both ad hoc and planned tasks. Normally, a clear picture of how all this will unfold does not emerge until fairly late in execution. More is said about this in Chap. 11.

Practicalities

Box 10.1: Two Fundamental Approaches to Estimation: Bottom-up and Top-down

The core parameters covered by a project plan involve estimates of a project's duration, outlays on external resources and demands for internal labour (and possibly other specific resources). It should be noted that there are two broad approaches to estimating these parameters: bottom-up and top-down.

Bottom-up estimation, also called work-based estimation, involves analysing each task (at the "bottom" of the WBS) in terms of: duration, outlays on external resources and demands on internal labour. These values are then aggregated to the overall project level. In this approach, we ask three questions about each task (the lowest level of the WBS):

1. Outlays on external resources. What materials and skills do we need to purchase for this task, and how much money must we outlay for those purchases?
2. Demand for internal labour. Who from the ranks of existing staff needs to be involved in this task, and how much labour (in person-hours) is required of them?
3. Duration. Given resource assumptions and availability, how long will it take to complete the task?

Top-down estimation, typically involves analysing each output and, based on the experience with similar outputs produced in past projects, directly generating estimates of labour, outlays and duration for that output.

Consider a project to build a wharf for a shiploader that requires 250 lineal metres of sheet piling. Outlays on material and labour, as well as the time involved would be based on unit coefficients derived from past projects (adjusted for the peculiarities of the subject project). If, for example, it was found that on similar past projects costs and productivity ran at about $12,500/lineal metre and 7.5 lineal metre/day, respectively, then a top-down estimate could be derived quite readily.

Bottom-up estimation is not always appropriate, especially when reliable estimates can be attached more readily to outputs than to the work of producing those outputs, as is often found in construction and engineering. In that case, a top-down approach appears to be more effective.

Because, in most projects, the original business case includes only a high-level WBS, the estimates of labour, outlays and duration provided there will have been derived from a top-down approach.

It should also be noted that a formidable range of estimation tools has been assembled over the past two or three decades; however, discussion of these lies beyond the scope of this book.

10.1.2 The Role of the Project Plan

The project plan provides the foundation for everything that is to be done during project execution. Zwikael and Sadeh (2007) found that planning reduces risk level and enhances success. At the end of planning, the baseline for project execution is made up of:

- The project plan (in its entirety).
- The project definition which appears early in the business case (the rest of the business case has been overtaken by the project plan).

10.1.3 Key Players in Planning

At the end of Initiation, the funder will have appointed a project owner. The project owner, in conjunction with the funder, then appoints a project manager. Section 7 of the business case discusses planning in broad detail—including some indication of the people who will be involved. Based on that, the funder, project owner and project manager finalise the core team to undertake the planning.

The bulk of the actual planning work falls to the project manager in consultation with the project owner. Throughout planning, if needed, both will meet with the funder. The level, form and frequency of contact between the project owner, project manager and funder will be dictated by accepted business practice in the funding organisation.

A project assurance counsellor (if appointed) has an important role to play in planning by:

- Liaising closely with the project manager and project owner (in a consultative capacity).
- Ensuring that the methodologies, processes and tools being employed are appropriate.
- Monitoring the quality of the overall plan.

10.1.4 Key Issues in Planning

Early in planning some important strategic issues must be considered. For example, How much of the work to be undertaken during execution should be contracted out? How is the scope of those contracts to be decided? Questions of that kind expose aspects of project management which tend to be more of a "black art" than a science. Dealing with such issues requires that the planning team has access to appropriately experienced people.

The quality of project plans varies widely. By "quality", here we mean the reliability and usefulness of the project plan as a script for execution. So, the central quality issue here with planning revolves around two questions:

- What are our estimates (of time and cost)?
- How reliable are those estimates?

A basic principle of statistics states that whenever an estimate of a parameter is provided—it must be qualified with an indication of its reliability. Some projects progress to execution with poor quality plans or, in some cases, with no plan at all. Inevitably, such a situation causes execution to degenerate into a collection of ad hoc tasks—with all the predictable consequences of increased costs, extended timeframes and reduced project worth. Not only should plans be assembled, but they should also be quality assured, by asking the question "How dependable is this plan as a script for execution"? Experience suggests that a depressingly large number of projects are undertaken without that question being asked—let alone answered.

10.2 Analysing the Work Involved in Execution

To plan project execution, we need to understand the nature of the process involved in producing the project's committed outputs—in particular, the dependencies amongst different parts of that process. Any of the (numerous) approaches to process modelling could be employed for this purpose, however, for a variety of reasons but the accepted and most effective techniques here are all based on the assembly of a Work Breakdown Structure (WBS).

10.2.1 The Work Breakdown Structure (WBS)

In, the ITO model (see Fig. 2.5), project execution is represented by the process ellipse that appears in the middle of the diagram. More formally, this ellipse could be generically labelled as something like: "build all committed outputs". Because it is so high level, such a process is clearly inadequate to guide execution because it sheds no light on the question "*How* as execution to be carried out?".

As the name suggests, a WBS takes the form of a structured (i.e. hierarchical) list of all the steps involved that work. Each output has a WBS (an output-level WBS). The (single) project level WBS is obtained by simply concatenating all the output-level WBSs. Because there is no "correct" WBS, the process of assembling one is largely exploratory—involving a sequence of somewhat tentative iterations.

While there is no conceptual bound on the number of levels that might be used in the structure of the WBS, as explained in Sect. 9.4.2, for many outputs, a *three*-level phase–activity–task (PAT) hierarchy is adequate. Combining all of the output-level WBSs results in a *four*-level hierarchy for the project overall: output, phase, activity and task. In accordance with the word-structure conventions introduced (see Sect. 2.3.3), in such a structure, all outputs are labelled with *nouns*, (because outputs are artefacts), and because they represent work of some kind, each item in the lists of activities and tasks is labelled as an imperative (an imperative is expressed as a directive, command or instruction to someone to carry out some action). While phases are also usefully expressed in this way, there will be occasions when some other form of labelling may be appropriate.

10.2.2 Hierarchical Decomposition

The technique of hierarchical decomposition is used to generate the WBS. Hierarchical decomposition is a simple but powerful method of converting a small number of large things into (an equivalent) large number of small things. In biology, the classification system of organisms is an example of a hierarchical decomposition.

To hierarchically decompose the work involved in project execution into a PAT-style WBS, we begin with the list of all the project's outputs, and then progressively break out the work demanded by each into three successively finer levels of detail.

While the number of outputs appearing in the first level of the WBS is set by the project scoping statement, the questions then arise:

- How many phases should be used for each output?
- How many activities should be used for each phase?
- How many tasks should be used for each activity?

The answer in each case is "7-ish" (a number of "close to" seven). "Sevenish" is suggested because of a trade-off between size of each list and the number of levels in the hierarchy. While a list of, say, two items is easy to remember, sort and analyse, shortlists demand more levels in the hierarchy (to cover all of the work to be done). Seven appears to be a useful practical rule of thumb.

A PAT-style WBS based on the rule of 7-ish would have about 350 tasks—which provides a reasonable level of detail for many, (but not all), outputs. If, however, each task is itself a relatively complex process, then it would be necessary to go beyond the task level. Clearly, as the number of levels in the WBS increases, labelling them becomes meaningless. It is for this reason that project scheduling tools call every item at all levels in the WBS a "task".

A WBS can get very large indeed. Extremely big projects—especially in defence—can have hundreds of thousands of tasks. For example, a WBS with 800,000 tasks will require about seven levels. In such cases, the task of creating and managing the WBS (as well as its attendant Gantt chart) is farmed out to sub-teams.

The elements in the lowest level of a WBS are, in taxonomic terms, called *primitives*. In project management, such a primitive is very usefully called a "work package". When does the process of hierarchical decomposition stop? There are two conditions under which work need not be broken up any further:

- When the current item can be delegated to a team member with no further instruction. An example would be "install light fittings". Because a qualified electrician knows how to install a light fitting, no further detail is required.
- When the current item takes the form of a natural process. Consider the task "allow nacelle (of wind turbine) to descend onto supporting column". The "lowering" is caused by gravity and so needs no further detail.

Two skills are needed to assemble a WBS:

- Someone with an understanding of the hierarchical decomposition process—such as a qualified project manager. Without such expertise, it becomes extremely difficult (if not impossible) to create a useful WBS.
- Someone with a general understanding of how each output is actually produced.

In summary, hierarchical decomposition starts with outputs and finishes with work packages. Except in the most trivial of cases, the assembly of a WBS (and accompanying Gantt chart) requires scheduling software. The process of assembling a WBS tends to be relatively ad hoc and exploratory with those involved skipping frequently between WBS levels and the Gantt chart.

As a matter of practicality, when incorporated into a Gantt chart, the final WBS must be displayed as a set of structured (indented) lists. During development, however, some people find other formats (such as block diagrams and mind maps) useful.

Illustrative Example

Box 10.2: A Project to Restore a Historic Homestead

The illustrations that are provided for various concepts throughout this chapter all relate to the following case study.

A family has just purchased a large historic homestead which has been classified by the National Trust. The building itself is in poor condition—as are the extensive surrounding grounds which, over many years, have become overgrown (and generally unusable) through lack of care by the previous owners. The family sees an opportunity of covering much of the cost of maintaining the property by opening it to fee-paying visitors at weekends.

A business case (prepared in support of the investment) includes the following elements of project scope:

A. Target outcomes:

- Increased quality of life for the family.
- New family income (from visitors).
- Increased value of family assets.

B. Committed outputs:

- A restored homestead.

- A landscaped garden.
- A shed for equipment, tools and supplies.
- Facilities for visitors (e.g. a kiosk, picnic areas and toilets).

C. Downstream processes:

- Regular family activity.
- Visits by guests.

10.2.3 *Developing a WBS for the Homestead Restoration Project*

The structure of a WBS can be illustrated using the landscaped garden output from the project described in Box 10.2. Typical phases for this work would be:

1. Engage a landscaper.
2. Design the garden.
3. Clear unwanted infrastructure.
4. Build new infrastructure.
5. Replant.

The *Replant* phase has, in turn, been broken out into the following five *activities*:

1. Purchase plants.
2. Dig planter holes.
3. Set plants in holes.
4. Update layout diagram.
5. Lay mulch.

Finally, the *Set plants in holes* activity have been broken into six *tasks*:

1. Loosen soil.
2. Set plant.
3. Backfill hole.
4. Stake plant.
5. Scatter fertiliser.
6. Water-in.

A number of points about this illustrative process are worth highlighting:

- It has been artificially constrained to one output, one phase and one activity (resulting in just six tasks). The same procedure would have to be carried out for all activities within all phases within all outputs.
- While none of these lists is exactly seven, they are "near" seven (and hence "7-ish").

- If the gardeners who had been contracted to perform the task *Set plants in holes* already knew how to carry out that work, then it would be unnecessary to break the *activity* into *tasks*—and so the activity becomes a work package.
- The lists of outputs, phases, activities and tasks will be peculiar to the project and the strategy adopted for its execution.

Notice the word structures used for the items in the above lists, they all take the form of imperatives (commands). An imperative is how an instruction is expressed for someone to complete an action of some kind. A principle applied to the construction of WBSs is that activities and tasks should be expressed as imperatives. There are exceptions to this rule, but it provides a powerful way of ensuring that work and the outputs from that work are not confused.

10.3 Developing a Schedule

When the work involved in execution is catalogued in a WBS, the duration of each task can now be estimated and the dependencies between tasks identified. Using these durations and dependencies, the WBS can now be displayed as a Gantt chart, which is a calendar-based diagram showing each task as a bar (positioned over a horizontal time axis) with:

- A start date (which locates the bar's left-hand edge).
- A duration (indicated by the length of the bar).
- A finish date (which locates the bar's right-hand edge).

10.3.1 The Gantt Chart

The detailed model represented by the WBS can now be employed to estimate the duration of the work involved in the production of all the project's outputs. This is obtained by estimating the duration of each task and noting any dependencies amongst those tasks. The process is called scheduling, the output from which is a timetable of some form.

In what follows we use three terms which, while commonly used in practice, are not always defined clearly: event, deadline and milestone. An event is an identifiable *instant* in time (it has zero duration). All elements of a WBS—in particular tasks—have duration and hence are not events. A task duration is gauged as the amount of time separating two specific events: when the work starts and when it finishes. The Gantt chart (discussed below) involves analysis of all the elements of a WBS, their durations and dependencies. Out of this analysis, an earliest feasible finish date can be derived for every task in the WBS. Because such finish events are supported with *evidence of achievability*, we call them *milestones*. *Deadlines, by way of contrast*, are set arbitrarily without any supporting *evidence of achievability*. Execution can be meaningfully tracked using milestones, but not using deadlines.

Scheduling allows a number of important questions to be answered, especially concerning any deadlines that may have been proposed:

- "Is the deadline feasible?"
- "If not, what must be done to make it feasible?"

The following major steps are performed to develop a Gantt chart:

1. Estimate duration. The elapsed time (for example, in days) is estimated for each task in the WBS. As explained in Sect. 10.3.2 below, the duration of a task is determined effectively by the allocation of resources.
2. Define dependencies among tasks. For example, in an office move project, the task "Unload the vans" cannot start until "Drive vans to new location" is finished. The effect of this is to constrain certain tasks by setting the earliest dates on which they can start.
3. Develop Gantt chart. There are two broad approaches to deciding the feasibility of a deadline: forward scheduling and backward scheduling. In forward scheduling, we begin by considering the start date of the project and work forward to calculate its end date, based on the durations of tasks/activities and dependencies among them. This procedure eventually reveals the earliest date by which the last task in the project can be completed. The feasibility of any proposed project deadline is readily decided by noting whether or not it occurs *after* the estimated finish date of the last task (if the proposed deadline occurs *before* the estimated finish date, then the deadline is infeasible). In backward scheduling, we begin by accepting the proposed deadline as the project end date and working backward to the start. If the indicated start date has already passed, then the proposed deadline is not feasible.
4. Identify the critical path. The critical path is a chain of linked tasks that, if not completed on time, will necessarily delay the finish date of a project (by the same amount). Tasks on the critical path are identified as "critical tasks". The critical path can be easily identified using most common project scheduling software packages. In addition to milestones, other tasks on the critical path should be monitored closely by the project manager during execution, because, as mentioned above, any delay with these will cause the whole project to finish later than planned.
5. Identify tasks with large slack. Slack (sometimes called "float") is the amount by which the duration of a task can be extended, without causing the project to be delayed (the critical path can also be viewed as the sequence of tasks with zero slack). Because such tasks are readily identified using a scheduling software package, project managers can use tasks with large slack to increase their flexibility in planning. For example, resources from such activities may be moved to activities on the critical path, if the result is shorter overall project duration.

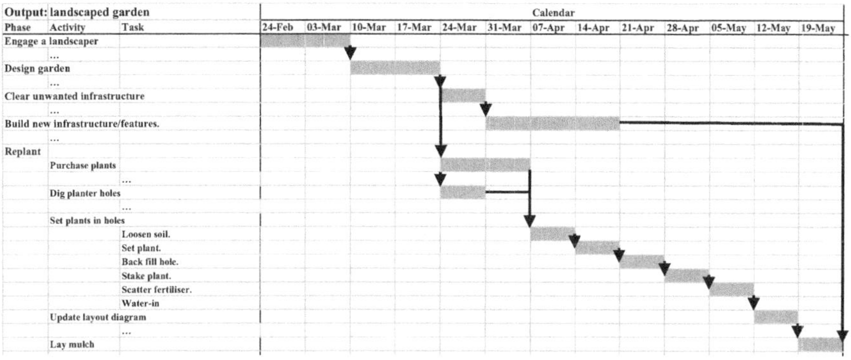

Fig. 10.1 A Gantt chart illustration

6. Address timeframe issues in the business case. Planning may reveal issues with the timeframe established in the approved business case. For example, it may now become clear that, given constraints on scope and budget, the original timeframe cannot be achieved. Even if the original timeframe was feasible, pressure may emerge during planning for an unrealistically short project. Before finalising a plan, project managers (in consultation with the project owner) are required to replace all deadlines with milestones. More is said of this in Sect. 10.3.2.

An example of a Gantt chart based on the information provided in Box 10.2 is presented in Fig. 10.1.

10.3.2 Deriving Estimates for Durations

Experienced estimators establish durations for tasks based on some notional allocation of resources. If at any time it becomes apparent that the assumed level of resource allocation is wrong, then they will adjust their estimates accordingly. The principles that underlie this practice are revealing, but, unfortunately, not always understood.

The duration of a task is dependent only on:

- The "intrinsic load" represented by the task in question. For example, assume the intrinsic load represented by the task "stake plant" in the garden renovation project is 750 plants.
- The notional rate of application of resources to the task, for example, "three gardeners".
- The proficiency of each resource. For example, each gardener is capable of staking 25 plant suppliers per day.

These variables can be related in a simple equation, where duration is equal to:

$$\text{Intrinsic\_load}/(\text{Rate\_of\_application\_of\_resources} * \text{Resource\_proficiency})$$

Using the illustrative figures above, it would therefore take 10 days (750/(3 * 25) = 10) to stake the plants.

In most real-life situations, the load from each task is given, as is the proficiency of available resources, and so the only discretionary variable normally available to the estimator which he/she can use to alter task durations is the level of resourcing.

Notice how in forward scheduling, *finish* dates are dependent on durations, while in backward scheduling, *start* dates are dependent on durations. Durations must never be calculated as the difference between two set dates. Inexperienced project managers, when faced with deadlines (especially if imposed by senior managers), automatically assume that their analysis is in some sense "wrong". In an attempt to "correct" the "error", they will go back through their work and change all of their estimates until the timeframe is now consistent with the imposed deadline. If the original estimated timeframe was accurate, any adjustments of that kind will, of course, lower the reliability of the project plan, and they cannot possibly improve its quality, and so are professionally unacceptable.

The achievability of a timeframe is decided by the quality of the schedule, not by the urgency of deadlines, and nor by the level of optimism reflected in the schedule.

10.3.3 The Schedule of Milestones

As explained in Chap. 11, monitoring project progress during the third global phase (execution) is effected by periodically comparing the *current actual* values of two particular parameters and the values of those same parameters that were *set* during planning. One of these relates to timeframe and the other to costs. The schedule of milestones takes the form of a simple table showing the dates by which, according to the Gantt chart, selected events must be achieved if the target date for completion of execution is to be achieved.

A milestone is an *event* (an instant) with two attributes, a *label* (identifying the event itself) and a *date*. While all sorts of events might be attached to a project as it unfolds, only those related to the critical path are suitable for tracking the progress of execution. The events in questions are those which signal the end of the tasks which lie on the project's critical path. By definition, a delay in completing a task on the critical path will delay execution by the same amount and so the ends of such tasks become the focus of our attention when judging the progress of execution.

A schedule of milestones takes the form of a simple table or chart that shows the end dates for selected milestones, and so two questions now arise:

- How many milestones are required?
- How are they selected?

Because milestones are used for tracking, there should be as many entries in the tables as there are review points for the project. If, for example, a 12-month project will be subjected to weekly reviews by the team, then the schedule of milestones will have about 50 entries. These need not be equally spaced, but the intention is not to have lengthy periods when no checkpoint on progress is available. As far as the selection of tasks to use as milestones goes, two simple rules of thumb are suggested:

- They should be scattered to occur "regularly".
- They should all be on the critical path.

Other milestones may also derive from contract commitments, for example, payment times, formal reviews and deadlines mentioned in the project contract.

10.3.4 Identifying and Managing Time Infeasibility

Consider the (all too common) situation when a project has had imposed on it an arbitrarily short deadline which is in conflict with that implied by the WBS and Gantt chart. If the Gantt chart has been based on reliable estimates of duration, then the imposed timeframe is provably infeasible—*given the assumptions underpinning the current analysis*. In that case, only three options are normally available to resolve the resulting time infeasibility:

1. Increase the rate at which extra resources are made available to selected tasks.
2. De-scope the project.
3. Relax the deadline.

We now examine each of these in turn and decide on their implications for the overall attractiveness of the project.

The application of extra resources inevitably involves higher costs, such as overtime or premium rates, with the result that reduced time usually means extra cost and hence lower worth. That is a trade-off that only the project owner can make (in consultation with the funder). While it is the responsibility of the project manager to identify and cost the options, he/she will normally have limited authority to make a decision that impacts the worth of the funder's investment. Whenever the "extra resource" option is exercised, its effects can be interpreted as a reduction in timeframe brought about by increasing project cost (although in reality the two are only indirectly related). Section 10.4.3 discusses this phenomenon in more detail.

Descoping the project involves either removing outputs or lowering the fitness-for-purpose of selected existing outputs. While such a move reduces both the timeframe and cost, because utilisation is compromised, it will also reduce the project's benefits. In general, these descoping effects, when all taken together, lower the worth of the project (if that were not true we would have to conclude that the project had been poorly scoped to begin with). Again, such an option can only be exercised by the project owner in consultation with the funder.

Finally, we have a third option, that of relaxing the deadline, which can have no impact on the project's:

- Timeframe (which is defined by the underlying milestones, not by desired deadlines),
- Cost (as distinct from its allocated budget) and
- Target outcomes.

And so, the project's estimated worth is left intact. Despite this (highly desirable) state-of-affairs, the stakeholder who proposed the deadline must, of course, agree to its relaxation (which may have political ramifications). In general, resolution of a time infeasibility will involve a combination of all three strategies.

10.4 Resource Planning

Much of planning is concerned with estimating the resources that project execution will require. This section offers a broad overview—with the emphasis on approaches that can be applied as they stand to smaller projects.

Over the years the project management discipline has developed very sophisticated frameworks and techniques to do this—detailed discussion of which lies beyond the scope of this book. Estimation in large projects—especially construction and engineering—is undertaken by specialists who are professionally recognised with titles such as *quantity surveyors* and *estimators*. They employ what we have described above as a top-down (*outputs*-based) approach—in which each project deliverable is analysed in terms of what products, materials, skills and work is required for its production.

10.4.1 Overview

The costs of a project are driven by the consumption of resources which in the ITO model are broadly grouped as labour and outlays on purchase of products and material. The estimated cost of a project should be distinguished from its budget. A project's budget is defined as the pool of money approved to cover project outlays. A project's cost is defined as the outlays required to acquire resources for the project. The term "budget" is frequently used to mean an arbitrary pool of funds that has been allocated to the project (often well before a business case is prepared), but for which there is no supporting analysis of resources. Because this use of budget is inconsistent with the definition used here, it can cause confusion, or even lead to a *cost-infeasible* project (see Sect. 10.5.4). Take, for example, the case of an organisation that has a history of budget "overruns" (because of unachievable budgets imposed on past business cases). In an attempt to pre-empt such a situation from arising on a new project, the funder may well set an artificially low budget as an (unstated) ambit claim, justified as being a "constraint". Because in that case, the budget has no supporting analysis or evidence of achievability, it cannot be used as a form of estimate.

The issue of whether to separate cash outlays from the notional value of internal labour involves some concepts that we do not cover here. For the moment, the immediate discussion applies equally to both purchased-in resources (for which cash will be required to make the necessary marginal outlays), and to internal labour (for which there will be some sort of notional headcount allocation or assignment of specific staff). Regardless of the treatment of labour, the project manager has a responsibility for producing reliable, verifiable estimates of cost, on which the project owner (in consultation with the funder) can then base a budget.

It is useful to identify any resource that will be specifically purchased as "external", while any existing resource which will be deployed without incurring any additional cash outlays as "internal". A resource must be treated as either external or internal—it cannot be both (which would result in double counting). This two-way classification gives rise to two resourcing plans for the project:

- The external resource acquisition plan—measured in monetary units.
- The internal resource deployment plan—measured as labour (typically in person-days).

Figure 7.1 introduces a three-way classification of project-related costs: production, management and (eventual) operations. The project budget is based on the first two of these (which relate to above-the-line and below-the-line works, respectively). The third is required for the financial analysis of the project. Planning will normally result in reasonably reliable estimates of all three and provide the foundation for the schedule of outlays in the project plan.

10.4.2 The Pattern of Planned Expenditure During Project Execution

The bulk of a project's costs is incurred during execution as resources are deployed (or purchased). A plot of the progressive cumulative total of planned expenditure against time is an important tool when making judgements about how project execution is progressing—as discussed in Sect. 11.2.2.1. We identify this as the planned expenditure (cost) graph—although it goes under a number of aliases. Various characteristics of the planned expenditure graph are worthy of note (refer to Fig. 10.2):

- Because it is cumulative, the planned expenditure graph rises monotonically and hence is often referred to as an "S" curve ("S" in this case stands for "sigmoid"—indicating that it often takes on a shape suggestive of a stretched letter "S").
- The x-values of the plotted points can be based on either a calendar (what costs have been incurred up until today's date) or on the project's schedule of milestones (what costs have been incurred up until the last achieved milestone).

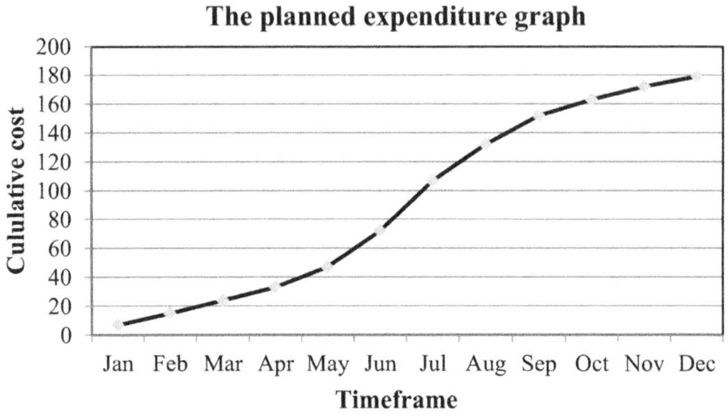

Fig. 10.2 A Project's planned expenditures—an illustration

10.4.3 The Relationship Between a Project's Cost and Its Duration

The relationship between project duration and cost appears somewhat counter-intuitive and so the matter bears some further discussion. In order to do that we have to consider the way that costs are driven by above-the-line and below-the-line works.

Below-the-line cost arises from the work to produce below-the-line outputs. Below-the-line outputs are those that appear in the project's statement of scope. Below-the-line cost and project duration behave as if they are inversely related, but the relationship is quite subtle. Neither cost nor duration can be manipulated directly, but both are sensitive to changes in a third variable which can be manipulated by the project manager. This is related to the rate of application of resources applied to tasks in the WBS. If premium (high cost) resources are withdrawn selectively from a task, in general, its duration will increase, and its cost will decrease. Consider, for example, what happens if we eliminate some agreed overtime to complete a task. Instead of gaining quick access to a skill (by paying someone to work longer out-of-hours), we must now wait for access to that same skill as it becomes available in normal working hours. The overall effect of this strategy is to lower cost (because normal hours are cheaper that overtime hours) but slow the rate at which the work is completed.

If, on the other hand, additional premium resources are selectively applied to a task, by reverse reasoning, its duration will decrease, and its cost will increase. These effects can be summarised as two "rules":

- Above a certain threshold, to reduce the duration of a task, selectively apply additional premium (high cost) resources. In general, this will increase its cost and so it appears that duration is falling in response to a cost increase.

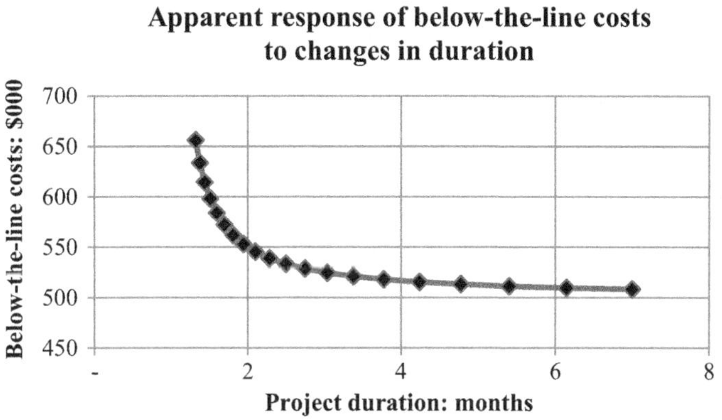

Fig. 10.3 The effect of resource manipulation on total below-the-line cost and duration

- Above a certain threshold, to reduce the cost of a task, selectively withdraw premium resources. In general, this will increase its duration and so it appears that cost is falling in response to an increase in duration.

In practice, numerous strategies are available to reduce duration, including the use of airfreight instead of surface shipping and working overtime—both of which are examples of high-price (premium) resources. Conversely, to reduce below-the-line costs, decisions might be taken to use surface shipping instead of airfreight and avoid overtime. The apparent relationship between below-the-line cost and duration—is shown in Fig. 10.3, where below-the-line costs are reduced by allowing duration to rise.

Above-the-line costs arise from above-the-line outputs (such as a monthly status report). These outputs are generic to most projects and often not included in the WBS (which is primarily concerned with the outputs included in the statement of scope). In some cases, these costs also include overheads (such as office rental and the fees for a part-time administrator). Because above-the-line costs arise from such activities as meetings, reports and the use of administrative tools, it is clear that they are driven primarily by the duration of the project (longer the project, the more meetings and reports that will be required). The behaviour of above-the-line costs in response to project duration is suggested in Fig. 10.4.

The total cost of a project is the sum of both types of costs, as is presented in Fig. 10.5. An appropriate choice of time/cost strategy will depend on the project's pressures and constraints at the time. For example, if total project costs have to be minimised, then duration may need to be extended (assuming no change in scope or risk exposure). If, however, the duration of the project must be minimised, then increases in costs have to be accepted.

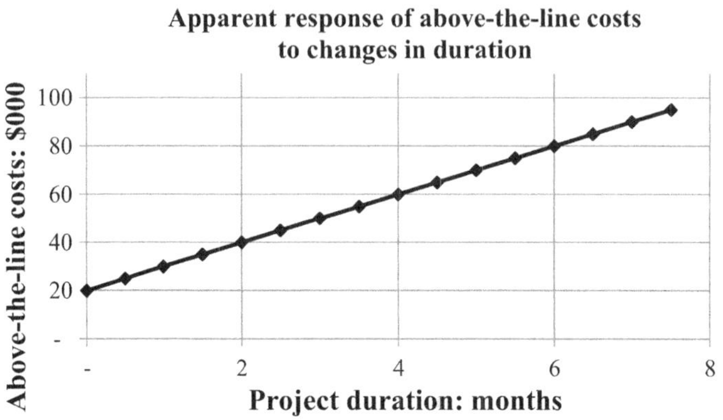

Fig. 10.4 Above-the-line costs responding to project duration

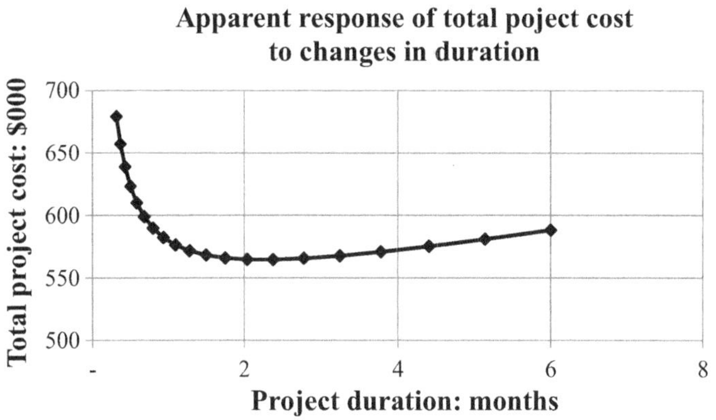

Fig. 10.5 Total project cost responding to project duration

10.5 Project Resourcing

Amongst other things, planning generates more reliable information on which the proposed investment may be reappraised before approval is given to begin execution. While the benefits sought by the funder from the exercise would normally remain as declared in the business case, estimates of other variables used in the investment decision—especially costs, disbenefits and riskiness—may well have changed. This is true not only for the *values* of these estimates, but also their *reliability*. We now turn our attention to the estimation of project cost—as reflected in the value of which are expected to be consumed during execution.

The approach and techniques used in planning are almost identical to those employed in relevant sections of the business case. Some of the variables covered in the business case may change during planning for either of two reasons:

- Planning goes into considerably more detail than initiation.
- As planning progresses, events unfold which may yield new information about what is to be expected during execution.

Our objective here is simply to explain to key players—particularly funders and project owners:

- What variables need to be estimated.
- How those estimates might be obtained.

Because we do not seek to explore estimation processes in depth, we base this discussion on a (somewhat simplistic) combination of the bottom-up (*work*-based) and top-down (outputs-based) approaches introduced in Box 10.1.

The ITO model accepts that inputs come from either of two sources:

- They take the form of existing resources, deployment of which requires no new outlays of cash. These are labelled as "labour" because they usually take the form of existing staff.
- They are purchased anew (with cash outlays)—loosely labelled collectively as "money".

In each case, the results of the analysis undertaken during planning takes the form of a table with columns related to time periods and rows to categories of resource.

For simplicity of discussion, we use point estimates (assuming that there is no uncertainty in the values of estimates), however, the reader should bear in mind the need to reflect uncertainty by using ranges instead of point values.

10.5.1 Internal Resource Deployment Plan

The internal resource deployment plan shows all those existing resources which are to be assigned to the project. This, in turn, reveals resources which will no longer be available for other (especially operational) activities during execution. Effectively, the internal resource deployment plan shows who is required when to do what and the level of intensity of that involvement.

Assembly of the internal resource deployment plan involves analysis of both the project scoping statement and the project schedule (represented by the WBS/Gantt chart). For every output appearing in the project scoping statement requiring the assignment of an (existing) internal resource, a list of assignments is assembled—based on a top-down analysis of:

- The timing of the assignment.
- The work to be undertaken by the internal resource during the assignment.
- The intensity of the assignment (typically in terms of person-days per period).

This analysis will not necessarily identify the total load imposed on internal resources. It could be that the project schedule demands, for example, regular coordi-

Table 10.1 Internal resource deployment plan (number of working hours)

Resource name	Feb	Mar	Apr	May	Jun	Jul	Aug	Sep	Total
Fred Nurke	15	20	25	35	50	40	30	30	245
Auger	0	0	0	0	10	20	35	35	100

nation meetings with suppliers—and so the same three-part analysis must be repeated (bottom-up) for the WBS/Gantt chart.

Because there are (by definition) no cash outlays for internal resources, the "natural" unit of measure is typically something like *person-days*. To illustrate this discussion with an example, assume that the owner of the property being landscaped (Fred Nurke) is a keen gardener and has decided to involve himself directly in some of the work. Furthermore, assume that he owns a tree-planting auger which is currently idle, but which he can make available for the project. This implies that there are just two internal resources—associated with Fred's time and the auger, respectively. All this can be summarised in a table such as that shown in Table 10.1.

10.5.2 External Resource Acquisition Plan

The external resource acquisition plan shows all those resources (products, materials, skills) that must be purchased (from others) for the project. This includes any resources that must be acquired to "back fill" all internal assignments. Effectively, the external resource acquisition plan shows: what must be purchased, when those purchases are to be made, the quantities required and the anticipated purchase price.

As was the case for the internal resource deployment plan, this involves analysis of both the project scoping statement and the project schedule (represented by the WBS/Gantt chart). For every output appearing in the project scoping statement requiring the acquisition of an external resource, a list of purchases is assembled.

Again, this approach will not necessarily identify the total demand for external resources, and so a bottom-up analysis must be undertaken for the WBS/Gantt chart.

Because (by definition) they all involve cash outlays, "natural" unit of measure for external resources is monetary (such as "dollars"). All this can be summarised in a table such as that shown in Table 10.2.

10.5.3 Costing and Budgeting for the Project

The completed external and internal resource plans now allow the project to be costed and a budget assembled.

The worth of a project is based on benefits, disbenefits and costs. While the project budget need to cover only the *outlays* represented by the external resource acquisition

Table 10.2 External resource acquisition plan (outlays in $000)

Resource	Feb	Mar	Apr	May	Jun	Jul	Aug	Sep	Total
Landscaping fees	8	10	8	2	2	2	2	6	40
New garden shed						5			5
Plumber			3	2	2	1	1		9
Electricians				3	4	2			9
Irrigation system			8						8
Plants							12		12
Concrete		3	6	7					16
Mulch							6		6
Excavator		4	4	4					12
Haulage		2	3	2					7
Disposal of spoil		1	2						3
Total	8	20	34	20	8	10	21	6	127

plan, we need to account for the *value* of the internal resources summarised in the internal resource deployment plan (despite the fact that they involve no marginal outlays). If internal staff are "backfilled", the costs of the substitute staff are already accounted for in the external resource acquisition plan and so the notional costs of the involvement represented in the internal resource deployment plan is zero. When they are not backfilled, a common practice is to value their work using payroll information for such details as salary with a notional allowance for on-costs (such as leave).

The external resource acquisition plan now provides the foundations or the project's cashflow analysis—such as that summarised in Table 9.7. There each column refers to a time period—for which total outlays have been estimated.

10.5.4 Identifying and Managing Cost Infeasibility

Section 10.3.4 explored the imposition of an arbitrarily short deadline which conflicts with that implied by the WBS and Gantt chart. A similar situation can arise if a budget is established which is arbitrarily smaller than that implied by analysis of the resources demanded by the project. Again, using a similar line of reasoning, four prominent options are available to resolve the resulting *cost* infeasibility:

1. Substitute non-premium-rate resources for whatever premium cost had been assumed.
2. De-scope the project.
3. Increase the budget.
4. Scale back any risk management plan that had been adopted.

The estimation process will have assumed a certain amount of premium-cost resources (of which zero is a special case). To the extent that such resources are replaced with non-premium (low cost) resources, then tasks will, in general, take longer to complete, but with reduced outlays. Dependant on the response of time-frames to reduced costs, such a substitution might well increase the worth of the project. As before, that is, a trade-off that only the project owner can approve (in consultation with the funder). While it is the responsibility of the project manager to identify and analyse the options, he/she will normally have limited authority to make a decision that impacts the worth of the funder's investment. Whenever the "reduced resource" option is exercised, its effects can be interpreted as an increase in timeframe brought about by reducing project cost.

As was the case for time infeasibility, descoping lowers the worth of the project because it constrains the utilisation of project outputs (and hence the execution of the project's downstream processes).

The third option (that of increasing the project budget) can have no impact on the project's:

- Timeframe (which is defined by the underlying assumptions about resources),
- Cost (as distinct from its allocated budget)
- Target outcomes.

And so, the project's estimated worth and its attractiveness are again left intact. As was the case with time infeasibility, the stakeholder who proposed the budget must, of course, agree to increasing the timeframe (especially in light of any political ramifications).

A decision to scale back any proposed risk management plan will have the effect of reducing the cost of the project—but causing at the same time an increase in riskiness. Such a trade-off can be made only by the project owner (possibly in consultation with the funder) In general, resolution of a cost infeasibility will involve a combination of all four strategies.

Practicalities

Box 10.3: Budgeting and the Handling of Uncertainty in Project Costs
Budgeting is the process of securing the pool of funds that will be made available to a project, based on reliable estimates. The processes of estimating and budgeting are independent (although clearly the feasibility of the project demands that the agreed budget exceed the project's overall estimated cost). A common confusion about the relationship between estimation and budgeting often causes high-risk projects to appear riskless. (Interestingly, this confusion is often propagated by the popularist treatments of projects seen in the press and electronic media.)

Take the example of a project with a high degree of uncertainty in its costs but with a budget that must be set as a fixed figure. While the uncertainty in costs is usefully expressed as a range, the demand for a fixed budget must not result in the range of cost estimates being replaced with an arbitrary point estimate, because point estimates imply certainty (i.e. a riskless project). The demand by a funder for a fixed budget in no way affects the project's level of uncertainty and so *estimates* must be provided as a range, regardless of *budgeting* requirements. It is a professional responsibility of the project manager not only to provide reliable information about costs, but also establish very clearly the level of certainty surrounding estimates of labour and outlays. *The project budget is informed by estimates, and estimates are not informed by the project budget.*

10.6 Packaging and Approving the Project Plan

When the material for the project plan is assembled, it is then packaged for formal approval by the project owner (where appropriate in consultation with the project funder).

As with the business case, packaging involves gathering together all this material, ensuring that is complete and internally consistent. In most cases, the tabling of the document would be followed by a detailed presentation and walkthrough.

Because the project plan and the project definition (declared in the business case) together serve as the baseline for execution they may need to be checked for consistency. The precise mechanism for appraising and approving the project plan will be determined by the policies and procedures of the funding organisation.

10.6.1 A Template for a Project Plan

Box 10.4 shows a typical structure for a project plan—which is explained below.

Template

Box 10.4: A Project Plan Template

1. **Introduction**

   1.1 Name of project

1.2 Purpose of project plan
1.3 Project overview
1.4 The project's key players
1.5 The relationship between the project plan and the business case

2. **Project Governance**
3. **Managing the External Project Environment**

3.1 Stakeholders
3.2 Risk
3.3 Issues

4. **Project Analysis**

4.1 WBS/Gantt chart
4.2 Schedule of milestones
4.3 Internal resource deployment plan
4.4 External resource acquisition plan
4.5 Project budget

5. **An overview of execution**

Here, we discuss each section of the template provided in Box 10.4:

1. **Introduction**: It provides a setting for the project plan.

1.1 **Name of project**: A project's name should clearly identify it and distinguish it from others in the funder's portfolio of projects.

1.2 **Purpose of project plan**: In general, this will be worded similarly for every project.

"*The purpose of this project plan is to:*

a. *Provide all the information required to approve a start on execution.*
b. *Along with Sect.* 2 *of the business case, provide a baseline for project execution.*
c. *Brief key stakeholders about the project*".

1.3 **Project overview**

- A concise overview of the project so that the reader gains a quick understanding of its overall "shape". The description should be succinct because the following sections provide all the necessary detail.
- A one-line appraisal summarising for the funder all three parameters that determine its worth: benefits, disbenefits and costs (by giving them each a value—such as *high, medium and low*).
- The wording of this section will depend on the audience, the extent of their background knowledge and the nature of the initiative itself.

1.4 **The project's key players**

Here, identify and briefly confirm the role of:

a. The funder(s),

   b. The project ownerand
   c. The project manager.

1.5 **The relationship between the project plan and the business case**
Discuss which sections of the business case remain relevant to project execution.

2. **Project governance**
This section defines the organisation model to be adopted during execution. It has two parts:

- The project governance model (as a diagram) supported with a brief description.
- A set of role definitions—one for each of the roles identified in the project governance model.

Because the flow of the business case would be interrupted if those roles definitions were inserted into this section, a hyperlink can be used instead, or they may be attached as an appendix.

3. **Managing the external project environment**
This section summarises the way in which the impacts of events arising from outside the project will be managed. The details provided here are ultimately drawn from the three registers: stakeholders (see Sect. 5.3.1), risk (see Sect. 6.4.1) and issues (see Sect. 6.5.4.1). It is impractical to incorporate them into the text of the business case for two reasons:

- They usually take the form of (relatively large) spreadsheets.
- Not all the information they contain is of significance for a funding decision.

To deal with these, for each of stakeholders, risks and issues three levels of detail are provided in this section:

3.1 **Stakeholders**

3.1.1 **Stakeholder management**

   a. A descriptive summary of critical information derived from the stakeholder register.
   b. A hyperlink to the stakeholder report.
   c. A hyperlink to the stakeholder register.

3.1.2 **Project communications strategy**

The project communications strategy takes the form of a table such as that shown in Table 9.9. If that is so big that it interrupts the flow of the document, then it may be attached as an appendix (with an appropriate reference inserted here).

3.2 **Risk**

   a. A descriptive summary of critical information derived from the risk register.
   b. A hyperlink to the risk report.
   c. A hyperlink to the risk register.

3.3 **Issues**

a. A descriptive summary of critical information derived from the issues register.
b. A hyperlink to the issues report.
c. A hyperlink to the issues register.

4. **Project analysis**

This section provides the results of a high-level analysis of the project definition

4.1 **WBS/Gantt chart**

In general, this will take the form of a workplan created with an appropriate scheduling tool (such as a project management software package).
This section will include:

a. A brief descriptive summary of any noteworthy highlights.
b. A hyperlink to the relevant scheduling file.

4.2 **Schedule of milestones**

This will take the form of a table—such as that suggested in Table 9.4. In general, it will be small enough to show here in full.
This section will include:

a. A brief descriptive summary of any noteworthy highlights.
b. The schedule of milestones itself.

4.3 **Internal resource deployment plan**

This will take the form of a table—such as that suggested in Table 9.5 (although in practice this would normally be more detailed—as would that for the external resource acquisition plan). This section will include:

a. A brief descriptive summary of any noteworthy highlights.
b. A hyperlink to the relevant internal resource deployment table.

4.4 **External resource acquisition plan**

This will take the form of a table—such as that suggested in Table 9.6.
This section will include:

a. A brief descriptive summary of any noteworthy highlights.
b. A hyperlink to the relevant external resource acquisition table.

4.5 **Project budget and cashflow planning**

This will take the form of a table—such as that suggested in Table 9.7.
This section will include:

a. A brief descriptive summary of any noteworthy highlights.
b. A hyperlink to the relevant cashflow table.

5. **An overview of execution**
   To confirm the project can progress to the third global phase (execution), this section will provide information about timing and resourcing implications by briefly summarising:
   - How long will execution take?
   - Who should be involved in execution? Doing what? When? How much of their time will be required?

10.6.2 Gauging the Quality of the Project Plan

Executing a project according to its plan does not necessarily ensure success. If the plan is flawed, the project is unlikely to generate its target outcomes. While a high-quality plan does not guarantee success, it increases the chances that the project will be properly executed and completed successfully. Conceptually, a high-quality plan is one that allows a reliable decision to be made about proceeding with the project. The reliability of decisions about a project is reflected in the eventual results of the exercise. The following discussion contrasts two approaches to measure the quality of planning.

Dvir and Lechler (2004) assess the quality of the project plan using six criteria:

1. The work breakdown structure is based on work packages at its lowest level.
2. Work packages have a target date for completion.
3. An estimate of the slack attached to every component of the work breakdown structure is available.
4. Dependencies amongst all work packages are known and specifically identified.
5. Reliable estimates of outlays for the project are available.
6. A resource plan is available showing the demand for all key personnel (who does what and when).

Zwikael and Globerson (2004) developed the Project Management Planning Quality (PMPQ) model to measure the quality of a project plan. The model builds on knowledge from the fields of project management, control theory, organisational maturity and top management support. Based on 33 processes that capture the planning processes conducted by project managers and top management support processes lead by executives, the model was used to evaluate the quality of project plans in different organisations. For example, they show that the quality of planning is higher in engineering and construction organisations but lower in government and production organisations. They also show empirical evidence that the quality of planning enhances in high-risk project scenarios as an effective risk management tool. Finally, Zwikael (2009) analysed the relative impact of PMBOK's knowledge areas during project planning.

Concerns about the quality of the estimates that underpin a plan can result in a rather untidy process which confuses *estimation* with *negotiation*. Take the case of a plan being appraised by a project owner, in which it is believed that the project manager has intentionally inflated durations and costs to increase the chances of meeting the agreed timeframe and budget. A response by the project owner in the form of an arbitrarily reduced timeframe and/or budget is a form of *negotiation*. Project success requires that estimates be, above all else, *reliable* (regardless of their attractiveness). Imposed arbitrary constraints will be, at worst, unreliable and at best of unknown reliability. Good practice suggests that, if the reliability of estimates is in question, they be assessed independently.

Summary

Planning seeks to assemble a script for the work required to produce the project's committed outputs. This script then allows estimates to be made about the timeframes and costs of execution. The analysis that underpins planning is a significantly more detailed and thorough version of that completed during initiation—with the objective of producing much more reliable estimates of timeframes and costs. The project plan, together with the project definition appearing in the business case, provides a baseline to project execution. Because, however, it is acknowledged that execution will not unfold exactly as described by the plan, a framework for eventually judging the progress of execution is also assembled—based on a schedule of milestones and resource plans. The next chapter discusses how the execution of a project is led and monitored.

Chapter 11
Executing a Project

Abstract Approval of the project plan is a significant decision because it implies that substantial work will now begin on execution which, in turn, requires the commitment of various resources. The processes that underpin initiation and planning are, in a sense, purely preparatory (because none of the work involved is related directly to the production of the project's committed outputs). By way of contrast, the bulk of the work undertaken during execution is concerned directly with the production, delivery and implementation of the project's outputs. The work of managing execution involves an iterative process *cycle*. This chapter describes this cycle and the tools used to support it.

11.1 Overview of Execution

Approval of the project plan is a significant decision because it implies that substantial work will now be undertaken to execute that plan. The purpose of execution is to produce, deliver and deploy the project's outputs. This global stage of a project requires the allocation of internal resources and, in most cases, the purchase of external resources. The processes that underpin initiation and planning are exclusively above-the-line (because none of the work involved is related directly to the production of the project's committed outputs). By way of contrast, the bulk of the work undertaken during execution is below-the-line (because it is here that the project's outputs are produced, delivered and implemented). Accordingly, project execution appears explicitly in the ITO model of the project—represented by the "process" ellipse in Fig. 11.1.

The management of execution also requires significant above-the-line work—in the form of an iterative three-part process cycle. The bulk of this chapter is devoted to a description of this cycle and the tools used to support it.

The execution global phase tends to dominate a project in terms of work, elapsed time and expenditure. Although the work of execution is undertaken according to the script represented by the project plan, evolving circumstances imply that the assumptions underpinning the initial script tend to become less relevant and realistic over time. This, in turn, implies that project execution tends to continuously "drift

O. Zwikael and J. R. Smyrk, *Project Management*,
https://doi.org/10.1007/978-3-030-03174-9_11

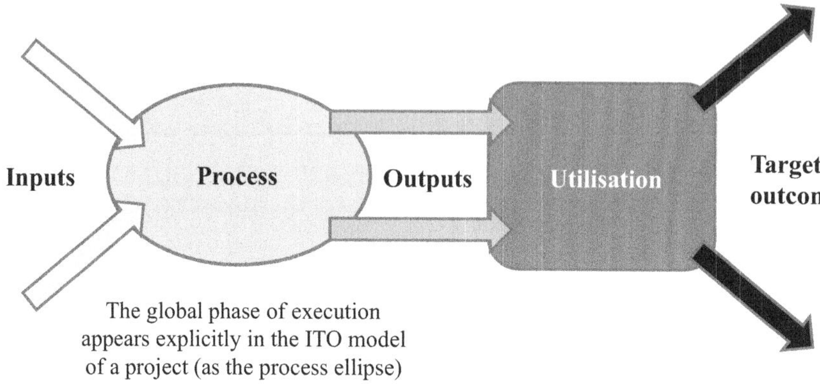

Fig. 11.1 The project execution global phase as an *element of the ITO model*

away" from the baseline adopted at the end of planning. At the same time, because the project is being undertaken as an investment, the underlying objective relates to the success of that investment. Project execution involves a control cycle directed at ensuring that the project and its baseline documents are always aligned. Project execution management takes place in regular forums of key players in which certain stylised status reports are considered for possible action.

The execution global phase can be usefully broken up into three (strictly sequential) sub-processes:

- Project setup
- Production of outputs
- Execution wind up.

The first of these relates to producing and implementing the list of *project setup outputs* which appear in the statement of project scope (see Sect. 9.3.2). Because these serve as infrastructure in support of execution overall, they will be scheduled for production at specific times—typically early in execution. Later, the project outputs themselves are developed by the project team. Finally, when all of the outputs that are in scope have been delivered, execution can be terminated, however, before doing that, some important housekeeping is necessary—usefully organised into a (third) concluding subprocess within execution. This subprocess "Wind up execution" involves formally taking down relevant project execution infrastructure and the disbandment of those elements of the project governance model that have no ongoing role in the next global phase (outcomes realisation).

As each committed output is produced and ready for delivery, it should be made the subject of accepted quality management practices including, where appropriate, quality assurance and quality control processes. Delivery of a committed output is a pseudo-commercial transaction involving the project's "supplier" (the project manager) and the project's "client" (the project owner). While it is the project owner who takes formal delivery of each committed output, physical delivery will normally involve the project customers (those who will utilise the outputs to generate target outcomes). Established commercial practice related to the fulfilment of a purchase order dictates that some form of declaration be made (or receipt issued) confirming that delivery has been affected. A signed delivery docket is an example of this. It is very desirable that those same commercial conventions be reflected in project practice.

11.2 The Project Execution Management Cycle

The project execution management (PEM) cycle is an *iterative* process made of three *sequential* subprocesses:

1. Project environment surveillance (PES).
2. Project execution control (PEC).
3. Project baseline revision (PBR).

Project environment surveillance involves monitoring the external environment to gain an understanding of events taking place in the surrounding world which might impact the way the project unfolds. Project execution control looks at gaps between the values of parameters set down in the plan and their current actual values. Using this information, actions are considered which might close these gaps. Project baseline revision attempts to deal with the situation in which it has become clear that no effective actions are available to close the gaps revealed in project execution control. Because all significant gaps of this kind must be closed, the only possible strategy involves revising the *intended* values of these parameters (which will, as a consequence, cause the worth of the project to change). For this reason, such revisions must be approved by the project owner (possibly in consultation with the funder).

The project execution management (PEM) cycle is the focus of two important regular forums—as discussed in Sect. 11.4 below.

11.2.1 Project Environment Surveillance (PES)

Project environment surveillance is concerned with the project's "surrounds". The three registers (stakeholders, risk and issues), together with associated summary reports, are the primary tools for this. Because of the way they are constructed, anything from outside that may impact the project will appear in (just one) one of the registers. Project environment surveillance is a process that seeks to answer three questions:

- What new events have been identified that could be relevant to the conduct of the project?
- What are the implications of these events for the way that execution will unfold?

- What action, if any, should be taken in response to these developments?

This process is carried out using summary reports on each of the three registers in turn. A report has exactly the same structure of columns as the associated register but is confined to entries that meet certain criteria as discussed below.

11.2.1.1 The Stakeholder Report

This takes the form of an extract from the stakeholder register which is confined to those entries (rows) that, *relative to the last report*, meet *any* of three conditions:

1. Have been the subject of significant change,
2. Are new and
3. Have been removed.

The very first such report shows stakeholders who (by definition) have not appeared in a previous report. Because all stakeholders are (as far as the first report goes) "new", the first *report* will be identical to the full *register*.

For all entries in the report, a decision is taken on how the stakeholder is to be engaged.

11.2.1.2 The Issues Report

This takes the form of an extract from the issue register that is confined to those entries (rows) meeting *either* of two conditions:

1. Have an importance rating of "High" or "Critical".
2. Have been assigned (or are proposed for assignment) for resolution to members of the steering committee.

For all entries in the report, a decision is taken on how the issue is to be managed.

11.2.1.3 The Risk Report

This takes the form of an extract from the risk register that is confined to those entries (rows) meeting *any* of four conditions:

1. Have a pre-expected damage value above a set threshold (see Sect. 6.4.2).
2. Have been realised since the last report.
3. Have been assigned to (or are proposed for assignment) to members of the steering committee for resolution.
4. Have been closed since the last report.

For all entries in the report, a decision is taken on how the risk is to be mitigated.

11.2.2 Project Execution Control (PEC)

Project execution control is concerned with ensuring that execution does not become detached from the project plan. The control mechanism takes the form of periodic reviews of progress against the project plan. This happens in two forums involving, respectively, the project team and the steering committee. Typically, project teams review progress every week, while steering commits meet every month. The focus of these meetings is on three questions:

- Is the project proceeding as planned?
- Will the plan be achieved?
- If the achievement of the plan is in doubt, what actions (if any) can be taken to correct the situation. In the event that no actions are available to ensure that the current plan is achieved, then the third part of the project execution management cycle (project baseline revision) is intended to support a decision about continuing with the project rather than terminate it.

The project plan includes, amongst other things, the two resource tables (internal and external). Both of these provide a projection of anticipated resource consumption (and hence cost) against time. The project's financial "state of health" is then judged against this projection—as discussed in what follows.

11.2.2.1 Why Planned and Actual Cumulative Expenditure Cannot Be Compared Directly

During execution, organisations are interested in two variants of project cost as suggested by Fig. 11.2:

- Planned cost—obtained as the progressive sum of all planned costs. The planned cost graph is complete.
- Actual cost—obtained as the progressive sum of all actual costs. During execution, the actual cost graph is incomplete.

A common method of judging the progress of execution is to compare the planned cost with the actual cost.

Because it provides no information about the progress of work required in the project, such a naive comparison reveals nothing about the project's state of health. This issue is revealed by noting that two contrasting project situations could both yield a graph along the lines of that shown in Fig. 11.2. The two project scenarios "A" and "B" are identical in all respects except that:

- In "A", project execution is now complete. All milestones scheduled for achievement over the entire project have been achieved (well ahead of time and at a lower cost than planned). Despite this rapid "burst" of outlays over the period since execution started, the actual expenditure on "A" sits below the planned expenditure

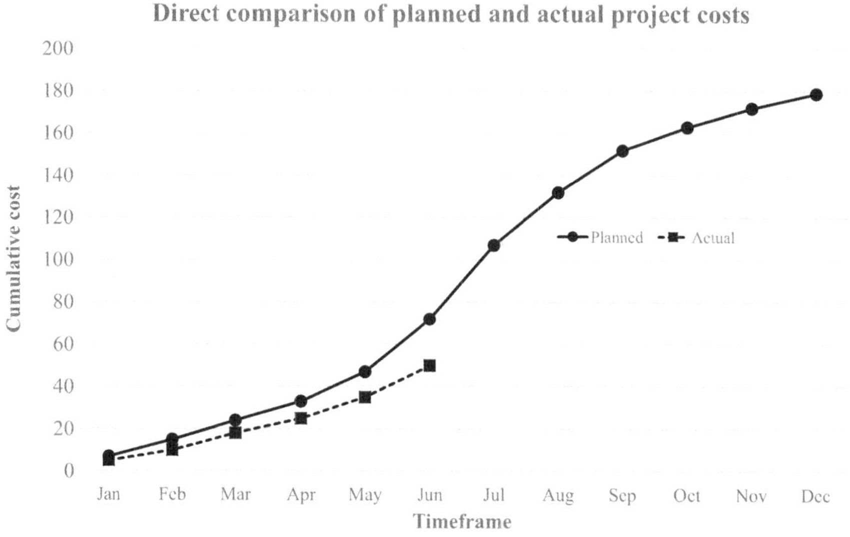

Fig. 11.2 A misleading tool for the analysis of project cost

graph. (Because not all of the money allowed for in the plan was consumed in acquiring the project's resources).

- In "B", few of milestones scheduled for achievement so far have been achieved. This lack of progress has resulted in less money than expected being spent to date. This "under-expenditure" causes the actual expenditure graph to fall below the planned expenditure graph—creating the illusion of underspending, whereas the real reason is lack of progress.

In summary, a simplistic comparison of the actual and planned expenditure is incapable of distinguishing a highly desirable state of affairs from a most undesirable situation, because it ignores information about what has been achieved in the project to date. Such analysis cannot, therefore, be used to judge adherence to the original plan. A number of approaches are available to deal with this issue. The most common one is to use the concept of earned value (EV)—as discussed below.

11.2.2.2 Using Earned Value to Judge a Project's State of Health

While a direct comparison of planned and actual expenditures reveals nothing about the status of the project, a comparison of each with a third variable does allow meaningful judgements to be made about adherence to the planned timeframe and budget. This variable is called "earned value" and is obtained by simply plotting the cumulative value of the work that has been completed (regardless of whether it is running early or late). The name "earned value" means that the work already completed in the project to date (e.g. completed outputs) has been earned by the project team and the monetary value of this work is counted in their favour. This method is commonly implemented by many organisations around the world and is a built-in component of most project management software packages.

Earned value is usually estimated at the activity level. The earned value of work performed is found by multiplying the estimated percentage completion of each activity by the cost for that activity. Comparing the earned value with the other two variables (planned work and actual cost) allows the identification of time and cost variances, calculating the efficiency of time and cost management, as well as predicting the final cost of the project based on the efficiency of the work to date (Zwikael et al. 2000). An extension to earned value (called "earned schedule") allows a more accurate forecast of the project's final duration (Lipke et al. 2009).

11.2.3 Project Baseline Revision

It is possible that the decisions and actions taken during project execution control are incapable of "realigning" the project with its baseline documentation. If the project cannot be changed to become consistent with its baseline documentation, then baseline documentation must be changed so that it represents an achievable project. For example, if, for any reason, the project owner concludes that the project is "out of control", he/she is acknowledging that execution can no longer be managed effectively. In other words, it is now accepted that no corrective actions can make the project feasible. The only rational alternative is to set new parameters which make the project feasible once again.

Three strategies for revising in the baseline documentation can be considered to achieve consistency with the project itself—all of which will require the approval of the project owner (where appropriate in consultation with the funder):

- Increase resources. Increases in approved resources cause costs to rise, and, hence, worth (and return) to decrease.
- Increase the timeframe for execution.
- Reduce scope/quality.

A decision to approve additional resources must be supported with such details as: the nature of the extra resources, the extra cost—as well as the impact on timeframe, worth and return. Similarly, a decision to approve an extension to the timeframe for execution must be supported with such details as: a list of the activities being delayed and the resulting impacts on worth and return.

Chapter 9 discussed scope at two levels:

- Level 1: where the list of committed outputs is set.
- Level 2: which takes the form of a collection of lists of fitness for purpose features—each such list *defines* a particular committed output.

The scope of a project can, therefore, be decreased by removing committed outputs or fitness-for-purpose features from the definition of an output. The quality of an output is reduced whenever fitness-for purpose features are removed (or weakened). In both cases, the effect is to reduce the amount of work represented by the WBS which, in turn, causes reductions in both cost and timeframe. The ITO model reveals that removal of an output from the project definition will, however, normally, prevent downstream processes from being executed as intended—usually with a severe impact on one or more target outcomes (some of which it may longer be possible to generate). This may, in turn, have a catastrophic impact on the attractiveness of the project. By way of contrast, removal or weakening of the fitness-for-purpose features of an output will *tend to reduce the usability of that output*—rather than prevent the downstream process from being executed at all. Again, the net effect is a reduction in project attractiveness.

11.3 The Project Governance Model

The entire project governance model, as confirmed in the project plan, is exercised during execution and so it must be ready for implementation soon after the project plan is approved.

11.3.1 The Project Manager

The project manager tends to dominate execution. This is true even if, because the size of the project merits formation of a project control group (which the project manager leads). The role definition for a project manager will normally follow a generic template within a funding organisation—with accountabilities and responsibilities that reflect current practice as accepted by the project management profession.

The project manager (and, by implication, the project administrator) have considerable responsibility during execution. They must be comfortable operating proactively in an environment that continuously demands reaction. Because the project and its environment evolve relentlessly, part-time assignments become problematic very quickly. Running projects "off the side of the desk" (with its attendant lack of attention to detail) represents a significant threat to all but the most trivial of initiatives (and so should be recognised as a generic threat in every risk register). If project administration is continually competing with below-the-line work for resources, time and attention, then execution management will suffer, with potentially catastrophic consequences.

The project manager must ensure that adequate resources are devoted to each of the three sub-processes of execution management discussed above:

1. Dealing with changes in the project environment. By ensuring that the three registers (stakeholders, risk and issues) are maintained, that the associated management programmes are being executed, that relevant stakeholders have appropriate access to them and that reliable reports flow to key players.
2. Controlling deviations from plan. By ensuring that the foundation instruments of planning and execution (in particular: statement of scope, WBS, Gantt chart, schedules of milestones and budget summaries) are continuously updated, that they are reported accurately to key players, that problems are identified (and advised) early and that recommendations for intervention are formulated quickly.
3. Revising baseline documents. By keeping the project owner fully informed of emerging pressures to revise the business case/project plan and providing reliable information on which any necessary changes to project strategy can be based.

11.3.2 The Project Owner

As agent of the funder, the project owner holds a particularly important accountability—that of achieving the return declared in the business case. All other roles recognised in the project governance model should be viewed as supporting that objective.

While good working relationships between the project owner and all those involved in guiding and managing execution are important to project success, two are particularly so: with the project manager and the project assurance counsellor (if one appointed), respectively. Under normal circumstances, these three will be getting together in frequent (often ad hoc) meetings (either as a pair, or together)—effectively serving as the project's "inner cabinet". These three must be comfortable with everything being tabled or discussed at the periodic meetings of the steering committee.

11.3.3 The Steering Committee

The project steering committee guides execution. Periodic project review meetings by the steering committee are conducted to ensure that, at all times, the project remains "pointed at" its baseline documents. Each meeting should be minuted—to record all substantive decisions and actions. This means that members are involved in a continuous cycle of review, analysis and, most importantly, *decision-making*.

Practicalities

Box 11.1: Enhancing the Effectiveness of Steering Committees

Anecdotal evidence suggests that, in general, steering committees do not meet as often as their role demands. One common reason given for this is that the same senior people are on all project steering committees and so they are heavily constrained in the amount of time they can give each project. Such a situation can reduce the effectiveness of a steering committee through effects such as:

- Irregular (or missed) meetings.
- Loss of continuity (amongst steering committee members).
- Poor levels of understanding amongst steering committee members about the status of the project.
- Inconsistency in the guidance and instructions from the steering committee to the project manager.

A partly related issue concerns the levels of skill amongst steering committee members to fill their role. Because they are often senior, experienced staff who have achieved their positions through a history of successful operational performance, it is assumed that they are automatically qualified to take on their project responsibilities. In practice, it appears that such stakeholders are encumbered by misconceptions about what a steering committee does, lack of awareness of their own roles and poor understanding of how projects should be undertaken.

All this exposes projects to unnecessary risks, which demands a well-considered response. Organisations are encouraged to formulate supportive policies that (amongst other things) might state positions on:

- The importance of steering committees.
- Meeting the executive load from steering committees.
- The obligations of members.
- Judging steering committee performance.
- Competency and capability standards for members.

11.3.4 Reference Groups, Advisers and Counsellors

The role description for reference groups and advisers includes details of their tenure. A sunset clause signals that, unless the role is extended, it will terminate on a particular date. During the period of their appointment, the role of each is declared explicitly in the associated role definition.

Where counsellors (project assurance and probity) have been appointed, their role continues beyond execution through to the end of outcomes realisation.

11.4 The Forums for Project Execution Management

The management of project execution involves two forums: one for the steering committee, the other for the project team (centred on either the project manager or the project control group). In each case, these forums review project status (based on standardised reports discussed in what follows), decide on actions and monitor the progress of earlier actions. Project team meetings would typically be conducted early each week—looking in particular at that part of execution scheduled for the following week. Although they have an agenda (and are minuted), project team meetings tend to be relatively short and informal. For example, status reports may be verbal; Steering committee meetings, on the other hand, would typically be conducted each month—supported with documented reports.

11.4.1 A Stylised Reporting Package

In steering committee meetings, the project manager reports on the progress of the project. Clearly, the project manager and project owner must liaise closely between meetings. It is important that they both support the thrust of the report and so one would expect that normally, the contents of each regular report would be agreed between them before it is tabled.

The project execution report has three sections, corresponding to the three subprocesses described in the discussion of project execution management provided in Sect. 11.2, with a structure along the lines of that suggested in Box 11.2. It is important that regular reports be kept simple and short, without any superfluous detail. Long reports are undesirable for three reasons:

- They require long lead times not only so that steering committee members can read them, but also so that the team can write them.
- They are out of date by the time they are tabled (because of the long lead time).
- They are expensive because they involve excessive above-the-line resources.

Template

Box 11.2: A Project Execution Reporting Package

1. **Introduction**
   1.1. Purpose of report. Simple, stylised statement that appears in every report.
   1.2. Executive summary. Summarises major observations, conclusions, issues and items requiring decision or action.
   1.3. Structure of the report. Simple overview of how the report is structured—to assist newcomers to navigate the document.
2. **Project environment surveillance**
   This section of the report advises stakeholders about important aspects of the project environment, summarises (and seeks approval where appropriate) for recommended programmes of management.
3. **Project execution control**
   The control of time and budget involves the acknowledgement of three horizons:
   - Project to date, using actuals (historical data).
   - To project end, using forecasts (future data).
   - At project end, using projections (actual + forecasts).

   3.1. Timeframe. A project progress report, based on the current schedule of milestones, with four short sections.
   - Milestones due for achievement since the last report and confirmation (or otherwise) of their achievement.
   - Milestones due for achievement before the next report and confirmation (or otherwise) of their achievability.
   - Implications for the project overall of the current rates of progress.
   - Discussion of any significant deviation from the currently approved workplan and proposed strategies to manage those deviations.

   3.2. Budget. This section covers:
   - Outlays (for the purchase of external resources).
   - Labour. Discussion of any significant deviation from the currently approved budget and the proposed strategies to manage those deviations.

   3.3. Deliveries
   - Confirm all deliveries effected since the last report and table quality certificates for each.
   - Confirm all deliveries to be effected before the next report.

4. **Project baseline revision**
   This section restates the key parameters that are to be recognised from this point on in the project, highlighting any that require change and formally seeking approval to accept the changes. A table format is useful with one row for each parameter and columns for: last approved value, new approved value, rationale for the change. The parameters to be reaffirmed or changed are:

   - Target outcomes (with full definitions),
   - Outputs (with lists of fitness-for-purpose features),
   - Undesirable outcomes (with descriptions),
   - Budget (outlays and labour) and
   - Timeframe (as a schedule of milestones).

11.4.2 A Stylised Agenda

Project execution management forums are strictly above-the-line, that is, they are confined to discussion of the managerial aspects of a project. This implies that it cannot be assigned (nor should it undertake) any tasks associated with production, delivery or implementation of outputs.

This separation of role does, however, become rather blurred from time-to-time. Take the example of a project to redevelop an old integrated iron and steel plant site as a multipurpose port. Tenders have been received to undertake the extensive (and expensive) civil work required to deal with significant ground pollution from the original coke ovens. Analysis of these bids shows a wide variation in prices, timeframes and risks. It is appropriate that when the steering committee makes a decision, it will not only consider costs, timeframes and risks, but also the criteria being used to rank the options and the technical analysis carried out by the project team. The latter topics are, in reality below-the-line, but any attempt at preventing such a discussion would be unreasonable (and undesirable). In practice, the steering committee agenda should be made up only of above-the-line agenda items, but participants should accept that discussion will inevitably "stray" below-the-line. It is up to the project owner (or whoever is acting as Chair) to ensure that the forum does not degenerate into some sort of competing "project team".

A steering committee meeting agenda that is consistent with the project execution reporting template suggested in Box 11.2 is presented in Box 11.3. For each substantive agenda item, there are the usual three sub-items: report, discussion and decisions.

Template

Box 11.3: A suggested agenda for meetings of both the steering committee and the project team

1. **Project environment surveillance**
   1.1. Stakeholders.
   1.2. Issues.
   1.3. Risks.
2. **Project execution control**
   2.1. Timeframe
   - Review project progress report.
   - Decisions about strategies to manage deviations.

   2.2. Resources
   - Review project budget report.
   - Decisions about strategies to manage deviations from plan.

   2.3. Deliveries
   - Acknowledgement of recent deliveries.
   - Acknowledgement of the achievements of the team in effecting these deliveries.
3. **Project baseline revision**
   - Review of all baseline parameters.
   - Decisions about changes in baseline parameters.

11.5 Outputs Closeout

A formal closeout process is followed as outputs are delivered from the project. The objective of this process is to progressively improve the performance of future projects by enhancing the overall project management framework adopted by the organisation. In some organisations, this is referred to as a "lessons learned review". Because of the nature of the closeout process, a workshop format (involving all key players) is common.

If outputs are delivered progressively over an extended period of time (as in the case of a large project), then a succession of closeout workshops may be appropriate. Closeout is strictly above-the-line, concerned with judgements about the planning and management of the project. Closeout is quite distinct from a conventional post-implementation review (PIR) which has a below-the-line focus, concerned primarily with judgements about the operation and quality of project deliverables. Because of their outputs focus, PIRs are relevant only to project execution.

The closeout process is based on an adaption of the widely accepted methodologies that underpin the continuous improvement of operational processes. The incremental improvements in successive executions of an operational process exploit the fact that (in most case) such processes are repeated. Each intervention in an operational process improvement cycle is normally effected by "adjusting" the script for the process in a highly controlled procedure. (In effect, this is what is happening during the project execution management cycle.) Because by definition, each project has a unique script (i.e. it is not repeated), there is no mechanism by which the script for the next project can be enhanced based on the script of the last. Despite this, above-the-line processes are repeated—and hence lend themselves to progressive refinement.

Two forms of closeout are normally conducted during a project: the first (discussed above) is called *outputs* closeout. The second is applied when the fourth global phase (outcomes realisation) comes to an end. This is called outcomes closeout. The methodology in each case is the same—they differ only in the list of the *project performance areas* adopted. The concept of *project performance areas* is discussed in Chap. 12—as are both variants of the closeout methodology.

At the end of execution, based on whatever outputs closeouts that have been conducted, all the information is available to measure the performance of the project manager and declare the project as a management success (or otherwise).

Summary

The committed outputs from a project are produced, delivered and implemented during execution. Most projects are completely dominated by the work undertaken during execution (both above-the-line and below-the-line). Conventionally, projects conclude when execution is complete, though it is important to manage projects until the target outcomes are realised or at least secured. For this purpose, the next chapter explores the fourth (and last) of a project's global phases (outcomes realisation).

Chapter 12
Realising Outcomes from Projects

Abstract When project execution is finished, all committed outputs become available to project customers—who are then able to participate in the execution of one or more downstream processes. The execution of downstream processes takes place in an *operational* (as distinct from *project*) environment. All going well, this will lead to the generation of the target outcomes which drove the original funding decision. It is possible, however, that target outcomes may not be realised as anticipated. This overlay of *operational* risk arises from a myriad of identified and unidentified threats associated with the execution of the project's downstream processes. To manage this risk, we take an early portion of the operational environment and treat it as a fourth global phase of the project—called "outcomes realisation".

12.1 The Context for Outcomes Realisation

As discussed in Sect. 2.5.1, the weak causal relationship between the utilisation of a project's outputs and the generation of its target outcomes is reflected in a level of uncertainty about the way utilisation will unfold. It is entirely possible that target outcomes may not be realised as anticipated. This overlay of *operational* risk arises from a myriad of identified and unidentified threats associated with the execution of the project's downstream processes. To manage this, we take an early portion of the operational environment and treat it as a fourth global phase of the project—-called "outcomes realisation", where outputs are initially utilised. This extension of a project's life represents a significant difference in approach to that assumed in most conventional views of a project—where the project finishes when the outputs are delivered.

12.1.1 An Overview of Outcomes Realisation

Although outcomes realisation is firmly embedded in the organisation's operational environment, it is formally structured as a global phase of the project environment. In

O. Zwikael and J. R. Smyrk, *Project Management*,
https://doi.org/10.1007/978-3-030-03174-9_12

particular, because accountability for the realisation of the target outcomes declared in the business case still rests with the project owner, he/she directs the last global phase—until eventually handing over to the executive who has functional responsibility for routine business operations.

During the outcomes realisation global phase, the project's key players seek to facilitate the generation of target outcomes—in accordance with the business case. In addition to that work, an outcomes closeout report should be prepared. Furthermore, the end of the entire project is usually, but not always, signalled by a formal handover to a functional executive. (There may be no such handover in cases where the downstream process is limited to either a set number of executions or a fixed period of time.)

Outcomes closeout uses the same methodology adopted for the outputs closeout conducted at the end of execution. In this chapter, we discuss, in turn:

1. The facilitation of utilisation during outcomes realisation.
2. The closeout process (which applies to both execution and outcomes realisation).
3. Handover to operations and project closure.

12.1.2 The Duration of Outcomes Realisation

Section 2.5.3 explained that, depending on the nature of the project, the duration of the operational environment (or the number of times that a downstream process is destined for execution) may be known in advance. If the period of output utilisation or the anticipated count of executions is bounded in some way, then the entire (time-limited) operational environment may be best treated as outcomes realisation—in which case an ongoing operational environment will not be established and there will be no handover to a functional executive.

The question now arises "If the duration of the operational environment is indeterminate, how much of it should be used as the foundation for outcomes realisation?". The reader will recall that the definition of a target outcome (see Table 9.1) includes an attribute "Date for achievement". The latest of these dates appearing in all of the target outcome definition tables determines the end of outcomes realisation.

As discussed in Chap. 8, a judgement about project ownership success is based on the status of the project *as at the end of outcomes realisation*. Project investment success, on the other hand, is based on the status of the *project as at the time of the judgement*. In other words, it is conceivable that a judgement made at one point in time could conflict with one made at a different time, because of more recent information becomes available. In short, declarations of ownership success are not time-sensitive, whereas declarations of investment success are.

12.1.3 Roles and Responsibilities During Outcomes Realisation

Many of the roles recognised in the project governance model cease at the end of execution. The project governance model that then remains in place during outcomes realisation would typically include:

- Selected members of the steering committee.
- The project owner.
- The project administrator (if appointed earlier).
- Selected reference groups and advisers.
- Project assurance counsellor (if appointed earlier).
- Selected suppliers.
- The project manager.

The role of the project manager during outcomes realisation bears further discussion. The project manager is accountable for the delivery of committed outputs. This accountability is fulfilled (or not) at the end of execution—when all outputs have been delivered. While the *accountability* comes to an end at this time, the *role* of the project manager in general and the accompanying responsibilities in particular should continue into outcomes realisation. This is because outcomes realisation can involve significant work (both above-the-line and below-the-line) which must be planned and managed—preferably by the project manager.

12.2 Facilitating Outcomes Realisation

The facilitation of outcomes realisation is concerned with the control and continuous improvement of downstream processes to ensure that they generate the project's target outcomes. The general strategy for doing this is constrained by the availability of options to intervene in the execution of each downstream process, which depends, in turn, on whether the downstream process is exogenous or endogenous. Because an endogenous downstream process has emerged as a project output, it offers significantly greater opportunity for intervention than is the case with exogenous downstream processes.

12.2.1 The Downstream Process Improvement Cycle

Facilitation involves monitoring the execution of downstream processes and, then, where appropriate, intervening to enhance performance. We propose a process improvement cycle based on four steps: monitor, measure, judge and modify. This approach is a variant of the Plan-Do-Check-Act cycle.

Each downstream process is formally monitored by measuring the variables adopted as target outcomes. (The reader will recall that a target outcome always takes the form of a process performance metric.) The observed measure is compared with the target levels set in the associated target outcome definition. If the gap is large enough to merit attention, then the project team intervene by making modifications to either the downstream process or selected project outputs (or both). If the downstream process is exogenous, then little, if anything, can be done to modify it. In that case intervention is indirect—confined to the fitness-for-purpose features of the project's outputs. If the downstream process is endogenous, then, in addition to modifying the fitness-for-purpose features, the team can also modify the process script.

12.2.2 The Role of IUMs in the Process Improvement Cycle

The integrated utilisation maps (IUMs) discussed in Chap. 9 contains the script of downstream process—as well as the critical fitness-for-purpose features of all the outputs that are utilised during its execution. With each process improvement cycle, these details are updated—in accordance with the proposed modification. In the case of endogenous downstream processes, the IUM is also revised to reflect agreed changes in the process script.

If the changes in either fitness-for-purpose features or the process script are relatively minor, then the modifications can be made as ad hoc tasks. At the other extreme, if the changes are significant, then they may well need to be undertaken as part of a follow-on project.

12.2.3 Closing the Project

When target outcomes have been secured (refer to Sect. 8.2.3), outcomes realisation concludes with a formal handover of all committed outputs and endogenous processes to a designated business unit. It is desirable not only to formalise this transfer with formal documentation but to highlight it with some sort of ceremonial event.

12.3 Outcomes Closeout

Section 11.5 briefly introduced, but did not discuss, outputs closeout—which is effectively a review of how well the project has gone up to the end of execution. The underlying methodology is the same for both outputs and outcomes closeout. Here we discuss that methodology.

12.3.1 The Closeout Process

A workshop format is suggested for the closeout review process involving everyone who is able to make substantive observations on the conduct of the project. Output from the workshop is a simple, short, stylised report.

Rather than examine all aspects of the way in which the project was conducted, the proposed closeout methodology looks at only a relatively small number of areas which, collectively, key players see as being important determinants of how the exercise unfolded. We call these "project performance areas" (PPAs). The closeout methodology involves the following steps (all of which are discussed in Sect. 12.3.4):

1. Select PPAs for review. It is recommended that the number of PPAs be limited to about eight—so that the analysis is thorough.
2. Assign an overall score—indicating how well this PPA was handled in the current project.
3. Provide evidence in support of that score.
4. Identify the causes (both positive and negative) which it is believed influenced the score.
5. Recommend changes in practice which would cause those factors to work more favourably in future.

The methodology is the same for both closeouts—they differ only in their timing and in the lists of PPAs that are considered. Outputs closeout is conducted as the last significant activity during execution. In a small project (or one in which all outputs are delivered over a relatively short period), there may well be only one output closeout workshop, but in large projects (or those in which outputs are delivered over an extended period of time), there could be multiple output closeouts (each one concerned with a specific set of delivered outputs). By way of contrast, there will normally be only one outcomes closeout (conducted as the last significant activity in outcomes realisation).

12.3.2 Project Performance Areas

Almost any aspect of a project can be used to frame a performance area—which takes the form of a review question—for example: "How effective were our risk mitigation actions?". Some topics would be of interest to the evaluation of performance in most, if not all, projects—such as those associated with costs and timeframes. In Tables 12.1 and 12.2, we catalogue some common project performance areas for output and outcome closeout, respectively—and suggest how each might be expressed as a review question.

Project performance areas are a somewhat loosely structured collection of elements—taking the form of anything from processes ("How effective were our periodic project status meetings?"), through documents ("Was the coverage of the busi-

Table 12.1 Some generic performance areas for outputs closeout

Project performance area (PPA)	Performance focus question
Fulfilment of scope	Did we deliver all outputs declared in the statement of scope?
Output quality	Were all outputs delivered in accordance with their definitions?
Project plan	How clearly did the plan set direction for the work of the project?
Internal resource estimates	Did the demands on staff accord with our labour estimates?
Outlay estimates	How reliable did our estimates of cost prove to be?
Risk management	How well did we manage the risks we faced?
Issues management	How well did we manage matters of concern that arose?
Timeframe	How well did we adhere to the timeframe approved in the business case?
Budget	How well did we adhere to the budget approved in the business case?
Governance	How well did the governance model—and in particular the Reference Groups?
Tracking	How closely were we able to monitor the project and influence developments?
Reporting	How useful were the progress reports tabled during the project?

Table 12.2 Some generic performance areas for outcomes closeout

Project performance area (PPA)	Performance focus question
Target outcomes	Were all target outcomes realised in accordance with the approved business case?
Undesirable outcomes	Did we exceed the notional thresholds for the undesirable outcomes identified in the business case?
Business case	How clear was it—and how reliable did it prove to be?
Disbenefits	Did the undesirable outcomes from the project accord with expectations?

ness case adequate?") to certain roles (Did we experience any breaches of probity guidelines?").

12.3.3 Preparing for a Closeout Workshop

Before each closeout workshop, the key players (especially the funder, project owner, project manager, project administrator, and counsellors) collectively agree on the

PPAs to be reviewed. The lists of generic PPAs provided above (Tables 12.1 and 12.2) offer a starting point—but others not shown there may well be included.

The same key players can now decide on a list of workshop participants.

Each review is, effectively, based on a comparison of what happened in the project and what was intended to happen. Accordingly, workshop participants will require access to the current baseline documentation and the most recent status reports.

12.3.4 The Closeout Report

In what follows we offer a template for a closeout report and provide an example (from a hypothetical project "Phoenix"). The closeout report has two parts: an overall (descriptive) summary, supported with results from an analysis of all the PPAs adopted for the review. Box 12.1 and its supporting detail suggest a template for a closeout report.

Template

Box 12.1: A Project Closeout Report

1. **Name of project**. The "official" name of the exercise.
2. **Purpose (of the report)**. Why has this report been prepared?
3. **Summary**. The main conclusions of the review.
4. **PPA analysis**: This section is made up of one-page summaries of the PPAs selected for review.

Some observations about this template:

1. **Name of project**: This should either repeat the title adopted in the project proposal—or appear as a refined version of that.
2. **Purpose (of the report)**: Generic wording along the lines of: "This report provides the results of a review of our overall performance on the project" would be appropriate.
3. **Summary**: Based on the main conclusions of the review, covering such topics as
   - What worked well? Why?
   - What worked poorly? Why?
   - What are the major implications for the conduct of future projects?
   - What now needs to happen so that our performance on future projects can improve?

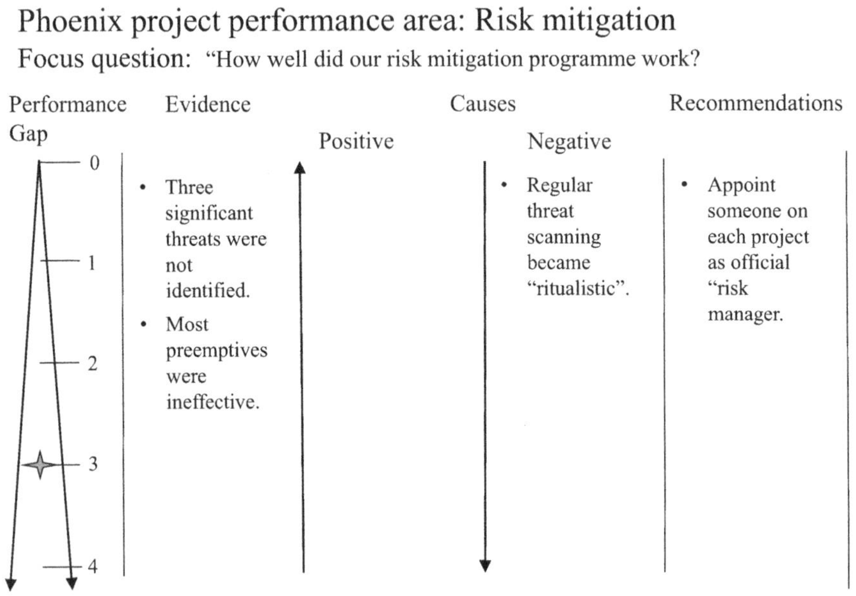

Fig. 12.1 An illustrative example of a PPA analysis report

4. **PPA analysis**: This section is made up of a collection of stylised one-page report-s—as described in the following discussion. We introduce the PPA analysis reports by providing an example from a hypothetical "Project Phoenix" (see Fig. 12.1) and explain its structure in Fig. 12.2.

Figure 12.2 shows how a PPA analysis report page is arranged—by indicating (with shaded rectangles) what details go where:

A. **Name of project**. The "official" name of the project.
B. **The name of the PPA**. This will be drawn from the list of PPAs adopted for analysis in the closeout process.
C. **The focus question attached to the PPA**. These correspond to the entries in the second column of Tables 12.1 and 12.2.
D. **The consensus score for the PPA**. Note that "0" implies perfect performance, while "4" corresponds to a very poor result. (The spread of the ladder legs suggests the gap between what was intended and what actually happened.)
E. **Items of evidence in support of the "gap" score**. The column beneath the heading "Evidence" is entered (as a set of bullet points) observations that support the score entered as "D".
F. **Positive "causes" of the score "D"**. In this column are entered (as bullet points) factors that worked in favour of the PPA.
G. **Negative "causes" of the score "D"**. In this column are entered (as bullet points) factors that worked against PPA.

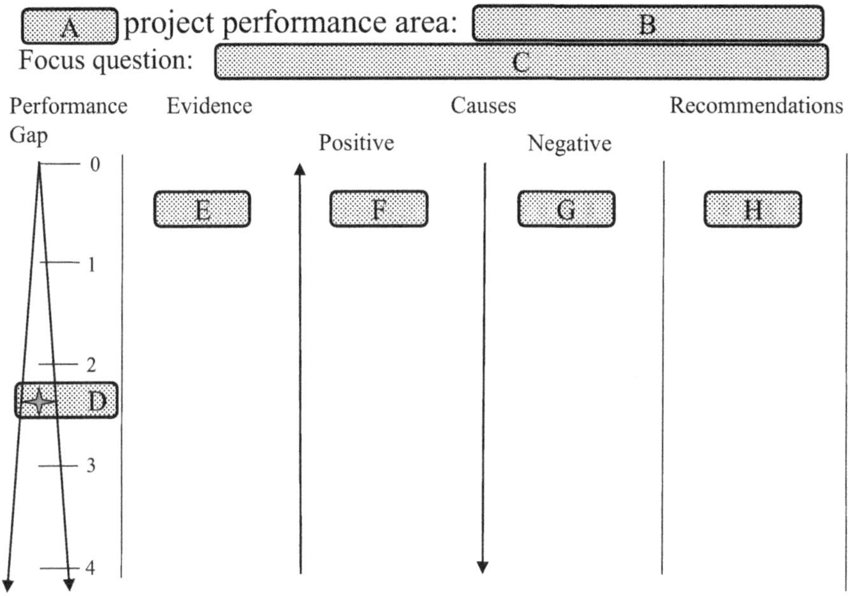

Fig. 12.2 The entries in a template of a PPA analysis report

H. **Recommendations**. In this column are entered (as bullet points) courses of action that would enhance the performance of this PPA in future projects.

Note how this approach is directed at performance enhancement through the *improvement of processes and procedures*. For this reason, it is undesirable to make individuals the subject of blame or praise (either directly or indirectly). Again, consistent with the philosophy of continuous improvement, the very name of the process ("closeout") is value-free—used in preference to other terms such as "lessons learned". It is important that the whole procedure be viewed positively. Terms such as "post mortem" are to be actively discouraged (because they suggest an underlying level of cynicism about the process).

Summary

The adoption of the ITO model as the underlying theoretical model of a project implies an *outcomes* focus. This approach implies an extension of a project's life beyond the delivery of outputs—into the securing of target outcomes. In this chapter, we have outlined how utilisation of a project's outputs can be facilitated—with the objective of realising the target outcomes appearing in the original business case.

Appendix A
An Integrated Glossary of Project Management Terms and Definitions

This glossary provides definitions of the core terms and concepts used throughout the book. It is an *integrated* glossary—in which:

- Each technical term has a specific meaning.
- There is a one-to-one relationship between these terms and the concepts to which they refer.
- Certain important terms used in the definitions are themselves defined elsewhere in the glossary.

A definition and/or a description (depending on the nature of the term) is provided in the second column for each entry. The definitions of certain terms include a "word structure"—a convention which dictates the way in which the labels attached to instances of the associated term are to be expressed. Where appropriate, discussion and/or examples are provided in the third column.

Term	Definition/description	Discussion/examples
2NY map	A diagrammatic representation of the three scenarios (Now/No/Yes) associated with a project funding decision	Used to assist in naming (and setting target values for) target outcomes The 2NY map shows values of the variables attached to target outcome under all three scenarios
Above-the-line (AtL)	Related to the administrative work on a project associated with its planning, monitoring, management and closure	Above-the-line is an adjectival expression. It is used to qualify both outputs and activities and refers to the administrative/management side of the project

(continued)

O. Zwikael and J. R. Smyrk, *Project Management*,
https://doi.org/10.1007/978-3-030-03174-9

(continued)

Term	Definition/description	Discussion/examples
		"Above-the-line", as a project management concept, should not be confused with the same term used in accounting, marketing and advertising Refer also: below-the-line
Accepted (project) practice	A collection of procedures that are judged by the project management profession as suitable for use in the management of projects	The guide to the project management body of knowledge, for example, includes a significant catalogue of accepted (project management) practices. These are frequently called "best practices"
Achieved business case	See business case (achieved)	
Activity	A formalised project process that is made up of tasks Also used loosely to mean any "block of work"	A stylised three-level work breakdown structure (identified as a "PAT-style" WBS), is used for discussion purposes involving: phases, activities and tasks Word structure: Activities are expressed as imperatives (command phrases) Refer also: phase, process, task, work package
Ad hoc tasks	Small "blocks of work" that require no formal planning or monitoring	The optimal way of handling ad hoc tasks is to "make them up as you go along"—correcting errors by simply redoing the work involved Isolated ad hoc tasks are usually not included in the WBS
Administrator	See project administrator	
Adviser	A provider of specialist input to the project who is not assigned to the project team	Advisers are part of the reference division in the project governance model The work of advisers is not directly supervised by the project manager An adviser is the name given to a "reference group of one" Refer also: reference group
Appraisal (of a project)	Assessment of a project carried out before funding is approved. (This is also called ex ante assessment)	The decisions to fund and approve a project are informed by an appraisal of the business case Any decision to abandon or continue a project in progress will be based on a revised appraisal of the current business case and/or project plan Refer also: assessment, evaluation

(continued)

(continued)

Term	Definition/description	Discussion/examples
Artefact	A (physical) output from a (human-directed) process	All outputs from (human-directed) processes qualify as artefacts Artefacts have a physical representation—with measurable physical attributes Refer also: output
Assessment (of a project)	A process to measure the attractiveness of a project	Takes two forms: • Appraisal—on which approval to proceed with the project is based • Evaluation—on which the quality of the (earlier) decision to proceed is gauged
Assessment test	A test that allows judgements of success or failure to be made about a project	A test is made up of: • A specific set of performance variables • A measurement on each of those variables • A set of criteria • A rule showing how to use the resulting measures to make a judgement about the project
Assessment trigger	An event that signals the need for a measurement of project-related performance	
Assumption	A condition placed on a project which, if not met, would require (potentially) significant revisions to the business case and confirmation of the funding decision	
Attractiveness (of a project)	An index that can be used to rank projects in terms of their desirability for investment Attractiveness is obtained by trading-off return against riskiness	The attractiveness of a proposed project determines how it ranks against competing ventures. This ranking can then be used by a funder to decide on when it will be funded (if at all)
Attractiveness map	A cartesian diagram with return on the Y-axis and riskiness on the X-axis	Used to rank projects according to their attractiveness for investment and to evaluate project success
Attribute	A defined and measurable characteristic	The word attribute is used here in a sense very similar to that adopted in data modelling In registers, entities take the form of instances (which are represented by rows), while columns are associated with attributes

(continued)

(continued)

Term	Definition/description	Discussion/examples
Baseline document	A document used to define, approve, monitor and evaluate a project	A baseline document is a comprehensive model of a project The two major baseline documents for a project are: the business case and the project plan
Baselining	The process of measuring the current (Now) values of target outcome variables	
Below-the-line BtL	Related the work of directly producing the outputs identified in the project's scoping statement	"Below-the-line" is an adjectival expression. It used to qualify both outputs and work Refer also: above-the-line
Beneficiary	A (project) beneficiary is a stakeholder who the funder wants to see "better off" as a result of the project	An entity who is better off because of the project, but whose gain is of no interest to the funder is not a beneficiary (in the formal sense). Such a stakeholder is a (positive) impactee Beneficiaries are not to be confused with a project's customers
Benefit	A notional flow of value (in the eyes of the funder) to a beneficiary —attributable to the achievement of desirable outcomes	A particular benefit stream will qualify for inclusion in the assessment of the project only if the funder believes it to be an appropriate "return" on the investment In general, benefits are gauged only indirectly (through their associated target outcomes)
Budget	The total value of the pool of economic resources approved for the project	The budget covers: • Approved cash outlays on purchased resources from outside the funding organisation • The opportunity cost of existing internal resources that have been made available to the project. (Usually in the form of staff deployed from participating organisations)
Business case	A baseline document that provides all the information required to make a reliable decision about funding a project	The business case is tabled by someone appointed by the funder as project champion, but usually assembled by an experienced project manager (under the direction of the champion). Following its approval by the funder, the business case is owned by the project owner

(continued)

(continued)

Term	Definition/description	Discussion/examples
Business case (achieved)	A notional version of the business case based on what actually transpired—rather than on what was intended to happen	The achieved business case contains only verifiable, relevant actual information. It is used to judge project success in the business case regression test
Business case (revised)	A revised and updated version of the (original) business case that was approved during the initiation phase	
Business process	A synonym for an operational process	
Champion	The person who seeks approval from the funder to assemble (or who is assigned responsibility by the funding organisation for the assembly of) a business case	The champion is the person or entity who drives the project through initiation to approval The champion is clearly a candidate to fill the role of project owner
Closeout	A process of judging a completed (or abandoned) project across a number of selected performance areas—with a view to improving the organisation's project capabilities	There are two classes of closure: • Outputs closeout: undertaken when selected outputs have been delivered—or the project abandoned. In large project, there may a succession of closeout reports • Outcomes closeout: undertaken when outcome flows have been secured
Closeout report	A report on the performance of the project across a number of selected performance areas	
Commissioned stakeholding	A form of stakeholding arising when an entity is appointed to fill a role in the governance model of a project	There are nine generic classes of commissioned stakeholder in a project: advisers, champion, counsellors, project manager, project owner, reference group members, steering committee members, suppliers, team members Refer also: spontaneous stakeholding
Communication strategy	A comprehensive list of all communications devices/channels to be employed in the project—broken out by stakeholder	
Constraint	A limitation on the conduct of a project imposed by circumstances	Constraints can take the form of either: an imposed value of some parameter, or a situation which cannot be avoided

(continued)

(continued)

Term	Definition/description	Discussion/examples
Contingency	A risk mitigating action that will have the effect of reducing the severity of the damage to a project that would arise from the realisation of a particular threat	
Contingency fund	A portion of the project budget set aside for either: • Undertaking any necessary contingency actions (as part of the project's risk mitigation programme) • Unforeseen issues	A contingency fund is not a "slush fund"—all draw-downs must be justified and documented
Cost(s)	The economic value of resources demanded by the project and the ongoing operation of its outputs	Costs arise from outlays of cash (to acquire products/services) and the deployment of internal staff Costs are of two kinds • Those incurred during the project itself (which may be further divided into management (AtL) and production (BtL) costs) • Those required for the ongoing operation of the project's outputs Costs and disbenefits are distinct concepts
Cost-infeasible (project)	A project for which the resource plan indicates a demand for more resources than those approved in the budget	The difference between the budget and the costs of the resource plan is a gauge of the level of (cost) infeasibility Cost infeasibility must be resolved before a project can be approved
Committed output	See deliverable	
Counsellor	See Project counsellor	
Criterion	A measure of performance to which a threshold value is attached	Refer also: performance variable
Critical path method (CPM)	A mathematical technique to determine the minimum duration of project execution—given the durations and dependencies of all phases, activities and tasks in the WBS	The critical path is represented by a sequence of tasks which share a characteristic—if they are delayed, project execution overall is delayed by the same amount
Deadline	A set date for an event in the project	Not to be confused with milestone. A project cannot be scheduled/tracked to deadlines—only to milestones Refer also: milestone
Defined output	An output from a project is defined when all its critical fit-for-purpose features have been identified and declared	Refer also: project scope, scope

(continued)

(continued)

Term	Definition/description	Discussion/examples
Deliver (a project output)	To transfer an output from the project environment into an operational environment—in readiness for utilisation	The accountability of the project manager is discharged on delivery Note that (because it is a process) a project is *executed* (not *delivered*) Refer also: produce, implement
Deliverable	An output that appears in a statement of scope	Deliverables represent a project's final outputs—as distinct from intermediate and above-the-line outputs Deliverables are also called "committed outputs"
Dependency	A logical linkage that establishes the execution sequence of two elements of a work breakdown structure (forcing a constraint on their timing)	There are four forms of dependency for pairs of activities: • S-S (start to start) • F-F (finish to finish) • F-S (finish to start) • S-F (start to finish)
Descope (a project)	To reduce the scope of a project	A project can be descoped in either/both of two ways • Remove outputs from the scoping statement • Remove (or weaken) the fitness-for-purpose features of selected outputs
Desirable outcome	An outcome that has value in the eyes of the project funder	Refer also: outcome
Detrimental outcome	An undesirable outcome attributable to a decision or action of the project manager with either of two characteristics: • It was anticipated, but subsequently experienced at higher levels • It was not anticipated, and subsequently experienced at unacceptable levels	
Disbeneficiary	A project stakeholder who experiences any loss of value from the project	A project's disbeneficiary is also a (negative) impactee
Disbenefit	A notional loss of value by an entity (attributable to the project)	A particular disbenefit stream may or may not qualify for inclusion in the calculation of a project's worth The loss of value associated with a disbenefit is always gauged from the perspective of the funder
Division (of a project governance model)	A project governance model has four organisational divisions (each related to the major types of role in a project)	The four divisions are: steering, delivery, reference and assurance

(continued)

(continued)

Term	Definition/description	Discussion/examples
Downstream (operational) process	An operational process for which performance metrics are identical to a project's target outcomes	Example: driving from point A to B following the completion of a new bridge—for which *travel time* is both a performance metric and a target outcome
Duration	The time taken to execute a process—in particular, any subset of a work breakdown structure (including: a task, an activity or a phase)	Elapsed time and labour are closely related, but distinct, concepts
Earned value management (EVM)	A project control technique which provides an analysis of a project's status and a prediction for its completion based on its planned work, actual cost to date and the work actually completed in the project to date	The methodology uses three key variables: planned value (PV), actual cost (AC) and earned value (EV)
Endogenous downstream process	A (new or re-engineered) downstream process	An endogenous downstream process is produced as an output from a project
Evaluation	Ex post assessment of a project—carried out after a project is executed	Evaluation allows a judgement to be made about whether the original funding decision is now cause for regret Refer also: appraisal, closeout
Event-impact model (of a risk)	The event-impact model of a risk has three (sequential) components: • A triggering event (threat) • A chain of consequences • A (damaging) impact	One complete instance of an event-impact model is called "a risk" The level of belief that a threat will occur is expressed in terms of likelihood (Li) The anticipated level of damage attributable to a threat is expressed in terms of severity (Se) The importance of a threat is expressed in terms of expected damage (ED) which is found as the product of Li and Se
Execution (project global phase)	The third global phase of a project—which is directed at creation of the project's outputs	Execution begins when a project plan is approved and ends when all the outputs in the statement of scope are produced, delivered and implemented (Or when a project is abandoned)
Exogenous downstream process	An existing downstream process execution of which is influenced in some way by outputs from the project	An exogenous downstream process is not produced as an output from the project
Failure	A state of unacceptable performance	Refer also: performance, success

(continued)

(continued)

Term	Definition/description	Discussion/examples
Fitness-for-purpose	Suitability for utilisation by a project customer	An output is defined when all its critical fitness-for-purpose features have been declared The quality of an output is expressed in terms of its fitness-for-purpose features
Fitness-for-purpose feature (FPF)	A characteristic of an output that makes it suitable for utilisation	A critical property of an output from a project
Flow of value	The stream of benefits (or disbenefits) that a stakeholder experiences (as valued by the funder)	Flows of value can be either: • Positive—giving rise to a gain in value (benefit); or • Negative—giving rise to a loss of value (disbenefit)
Forms of stakeholding	The stakeholding that an entity has in a project can be of two forms: • Spontaneous (arising from the nature of the project) • Commissioned (arising because of an appointment to fill an assigned role in the project)	We recognise six classes of spontaneous stakeholding in a project and nine classes of commissioned stakeholding
Fortuitous outcomes	Desirable outcomes that are not targeted in a project's business case	Fortuitous outcomes can play no part in the (ex ante) appraisal of a project. Realised fortuitous outcomes are, however, recognised in project evaluation when gauging the project's actual (*ex post*) worth
Funder	A stakeholder who commits funds and/or approves the allocation of labour to the project	A project can have multiple funders Note that the funder decides on funding. The issue of where the money originates (e.g. a loan from a bank) is irrelevant to the definition
Funding organisation	The organisation to whom the funder belongs. (A funding organisation owns the money for which the funder is responsible)	Refer: funder
Gantt chart	A diagrammatic representation of a project schedule	In a Gantt chart: • The vertical axis identifies elements of the work breakdown structure • The horizontal axis shows time (as a calendar) • Each element of the work breakdown structure is represented by a horizontal bar indicating start, duration and finish • Links between bars indicate precedence

(continued)

(continued)

Term	Definition/description	Discussion/examples
Global phase (of a project) model	The life of a project is made up of four global phases: • Initiation • Planning • Execution • Outcomes realisation	A global phase can be undertaken only of the one before has been completed. A project can be terminated at any point
Governance model	An organisational structure that recognises and integrates responsibilities for outcomes, outputs and work by defining, for a structure of entities, their roles and relationships	
Hierarchical decomposition	The recursive process of describing an element by breaking it into smaller elements	Effectively, hierarchical decomposition converts a small number of large concepts into a large number of small concepts
Impact	The last (of the three) components of a risk. An impact takes the form of damaging effects that are reflected in a lowering of the worth of a project	Table 6.2 summarises the six forms of impact that may arise from a risk
Impactee	A (project) impactee is a stakeholder who experiences either • A loss of value as a consequence of the project being undertaken—regardless of whether or not it achieves its target outcomes. (Negative impactees) • A gain in value from non-target outcomes (Positive impactees)	
Implement (an output)	The activity of preparing an output ready for utilisation by a project customer	
Infeasible (project)	A project for which parameters have been set that are internally inconsistent	In an infeasible project, no programme of work is available that would allow the parameters to be achieved—and so the project cannot succeed Infeasibility arises from constraints on either/both of • Costs/outlays/resources • Timeframe Refer also: cost-infeasible (project), time-infeasible (project)
Influencer	A (project) influencer is a stakeholder who, by virtue of his/her role, standing or position carries significant body of opinion about the project	Influencers are important targets of the project's communications plan because their views determine the views of a large number of stakeholders

(continued)

(continued)

Term	Definition/description	Discussion/examples
Initiation	The first global phase of a project —directed at assembly of a business case	Initiation is the first phase of every project
Input	An economic resource that is consumed during the execution of a process	In projects we classify inputs into two broad groups: • Labour • Purchased resources (which we broadly label as "money")
Input-process-output (IPO) model	A (long established) model that seeks to explain the relationships amongst: inputs, a process and outputs	The conventional view of a project corresponds to an IPO model of that project
Input-transform-outcome (ITO) model	A model that seeks to explain the relationships between outputs and outcomes	The ITO model was developed by John Smyrk in the late 80s. It extends the IPO model by incorporating a mechanism called utilisation
Integrated Utilisation Map (IUM)	A tool shows when each committed output is utilised during execution of a particular downstream process, and what characteristics that output must have if it is to be successfully utilised	There is one IUM for every downstream process Refer to Sect. 9.3.4.2
IPO	See input process output	
Iron triangle	A conventional test of success for project management (Sometimes called the "golden" triangle or the triple constraint)	The "iron triangle" requires that all the project's outputs: be delivered; in accordance with an agreed specification, on time and within budget Refer also: steel tetrahedron
Issue	An existing matter of general concern for the project (other than a risk) that requires resolution (or a response)	The management of an issue is distinct from that used for a risk. An issue can give rise to one or more risks
Issues life cycle	The particular sequence of states through which an issue passes between being identified and resolved	The states that an issue can assume in its life cycle are: identified, parked, prepared, active and resolved
Issues register	A tool for cataloguing issues and documenting their management	
Issues report	A report that highlights important issues for special consideration	An issues report is a subset (in terms of entries) of the issues register
Issues management	A (largely clerical) procedure for identifying, analysing, tracking and resolving issues	
ITO	See input-transform-outcome	

(continued)

(continued)

Term	Definition/description	Discussion/examples
Key stakeholders	Those stakeholders in the project who are included in the stakeholder engagement plan	
Likelihood (Li)	A qualitative measure of the level of belief (or confidence) that a threat will occur	Likelihood is similar to the concept of probability. Li has a value drawn from the closed interval [0, 1] The Li of a threat before an RMP is put into place is called "Pre-Li" The Li of a threat after an RMP is put into place is called "Post-Li"
Loss of value	See flow of value	
Memorandum of understanding (MoU)	A formalised agreement under which an internal entity releases resources to a project	MoUs lay out the terms under which the resource is made available and are somewhat akin to contracts—but normally have no legal standing
Methodology	A systematic, standardised high-level model of a generic process. Such a model is used as a starting point in the development of a comprehensive script for a particular application of that process	This term is used in two distinct ways in the project management literature: • To indicate a recognised process (a method). For example, there are established methodologies for undertaking social surveys • To indicate an entire framework of tools, techniques and templates for managing projects (in that sense PRINCE2 is a project management methodology)
Milestone	An event that has been adopted for use in tracking a project's progress. These are usually selected from those events that define the finish of tasks on a project's critical path	Milestones are derived analytically from the project's workplan and are not to be confused with deadlines (which are set arbitrarily)
Natural object	Something physical which was created by natural (as distinct from human-directed) processes	Refer also: artefact, null
Net present value (NPV)	A number representing the combined value (in today's terms) of a future monetary flow	
Network diagram	A diagrammatic representation of the activities included in a work breakdown structure, their durations and interdependencies	Such a diagram is known in mathematics as a "directed graph"
Null	A special instance of a project output in which an existing artefact or natural object is eliminated/destroyed/demolished	The destruction of Ripple rock (Box 2.8) resulted in a null (in the form of a "destroyed reef") Refer also: artefact, natural object

(continued)

(continued)

Term	Definition/description	Discussion/examples
Objective (statement of)	A succinct statement that explains why the project is being undertaken	The (project) objective statement is one of five elements that make up a project's statement of scope
Operand	An identifiable artefact or object which a process seeks to change physically	Usually in the form of either an existing artefact or a natural object
Operational environment	That part of the business environment within which regular (non-project) business activity is conducted	
Operational process	A regularised "block of work" for which there is an existing script	
Outcome	An indirect result (attributable to an identifiable process or mechanism) that takes the form of a measurable change in some variable that describes the "state of the world"	
Outcomes realisation	The fourth global phase of a project—directed at securing the flow of target outcomes	Outcomes realisation is usually only part of the operational environment that is created by a project Outcomes realisation begins when customers start utilising the project's outputs. It ends when the flow of target outcomes has been "secured"
Outcomes coverage matrix (OCM)	A tool used to analyse the robustness, completeness and appropriateness of a project's scoping statement	There is one OCM for a project. It takes the form of a matrix with entries of either "0" or "1". Columns are associated with target outcomes and rows with downstream processes A "1" means that the outcome at the top can be used as a performance metric for the downstream process at the left
Outlay	Expenditure of cash on external products/services demanded by the project's work breakdown structure	
Output	An artefact created during execution of a process	An artefact is a direct result of process execution. In project management, we are concerned only with outputs from human-directed processes Word structure: Outputs are expressed as nouns. Verbs are not used—nor is the word "to"

(continued)

(continued)

Term	Definition/description	Discussion/examples
Output classes	Outputs are classified according to the reason they have been included in the scope of the project	Refer to Sect. 9.3.2.3
Overengineering (of an output)	A situation in which the definition of an output includes superfluous fitness-for-purpose features)	An overengineered output incurs more costs and takes more time than is necessary Refer also: underengineering
Overscoping (of a project)	A situation in which the current scope of a project includes redundant outputs	An overscoped project incurs more costs and takes more time than is necessary Refer also: underscoping
Performance	A measure of achievement	Performance is *measured* (or gauged). By way of contrast, success is *judged* (or determined)
Performance variable	A variable used to gauge performance (on a project)	
Phase	The highest level of the stylised three-level PAT (phase, activity, task) work breakdown structure discussed in the text (See Sect. 9.4.2). Phases are typically sequential	For certain well-understood types of outputs (such as bridges, business processes and software) a standardised sequence of phases has been adopted as "accepted practice". Such a phase sequence is called a "methodology"—one of a number of different meanings attached to this word Refer also: WBS
Planning	A global phase of a project directed at preparing a detailed model of the work required during execution	Planning is undertaken by the project manager Planning begins when a funder accepts a business case. It ends when a plan is tabled for consideration by the project owner
PMBoK	(A Guide to) The project management body of knowledge	A compendium of accepted project practice developed and maintained by the PMI
PMI	The Project Management Institute	One of the most influential professional bodies in the field of project management. As well as producing the PMBoK, it also manages a framework of professional accreditation
Pre-emptive	A risk mitigating action that will have the effect of reducing the likelihood of a threat emerging	Refer also: contingency
Probity counsellor	A specialist independent adviser appointed by the steering committee to ensure that the commercial arrangements between the project team and the outside	Otherwise known as a probity adviser The probity counsellor is not an auditor—and so is free to work

(continued)

(continued)

Term	Definition/description	Discussion/examples
	world are conducted in accordance with accepted procurement practice	consultatively with the project manager and other key players
Process	Human-directed work that produces one or more outputs	
Process metric	A variable observed after one or more executions process that can be used to gauge the performance of the process	The metrics of a downstream process include one or more target outcome variables Three variables serve as generic process metrics: output count/volume, duration (of execution) and cost (of execution)
Produce (a project output)	The activity of assembling an output ready for delivery to a project customer	Production is the first of three steps associated with creation of an output (the other two being delivery and implementation) Refer also: deliver, implement
Programme evaluation and review technique (PERT)	A method of generation a probability distribution function for the duration of project execution	This method allows the probability of meeting any given end date to be decided
Programme (of projects)	A set of coordinated projects	Projects may be coordinated under either of two different scenarios: • One or other of the projects will suffer a reduction in its worth because one project "interferes" with the other • The current worth of one or other of the projects can be increased by electing to coordinate them
Programme manager	An individual who oversees multiple related projects and their project managers	
Project	Non-repeated planned work intended to enhance organisational performance	Not all processes are suited to being managed using a project management framework. In general, processes that are novel and complex and do not have a script for their execution are best managed as projects
Project administrator	Supports the project manager by overseeing the project's above-the-line work	Project administration is an above-the-line role—which reports to the project manager Project administrators are only required when the project's above-the-line work cannot be adequately resourced by the project manager (or selected team members). Large projects may have more than one project administrator

(continued)

(continued)

Term	Definition/description	Discussion/examples
Project assurance counsellor (PAC)	A specialist independent adviser appointed by the steering committee to ensure that the project is being conducted in accordance with whatever framework of management has been adopted by the funding organisation	The project assurance counsellor is not an auditor—and so is free to work consultatively with the project manager and other key players
Project baseline revision	The process used to change parameters set in the project's baseline documents	
Project cost	The value of the resources consumed by a project	Three forms of project cost are distinguished: • Project management costs: the value of resources consumed by a project's above-the-line work • Project execution costs: the value of resources consumed by the below-the-line work undertaken during execution • Total project costs: the sum of project management costs and project execution costs Note that project costs are confined to resources consumed during execution and outcomes realisation. The ongoing costs incurred by execution of downstream processes in the operational environment are excluded
Project counsellor	A specialist independent adviser appointed by the steering committee to ensure that the project is being conducted in accordance with accepted practice	Two project counsellor roles are recognised in the project governance model: • Project assurance counsellor • Probity counsellor
Project customer	An entity who, by utilising outputs, contributes to the generation of target outcomes	A project can have multiple customers
Project environment surveillance	The process of monitoring the project's environment—to ensure that external factors are appropriately managed	Project environment management is based heavily on the three registers: stakeholders, risk and issues
Project execution control	The process used to identify and manage deviations from plan during the execution phase of a project	Project execution control is applied in particular to: • Timeframes • Resources (outlays and/or labour) • Output quality

(continued)

(continued)

Term	Definition/description	Discussion/examples
Project execution management	The above-the-line work surrounding execution	Project execution management involves three kinds of process: • Project environment surveillance • Project execution control • Project baseline revision
Project management office (PMO)	A group that typically standardises the project-related governance processes and supports project managers within the same department, agency or business unit in areas such as project management methodologies, tools and techniques	There is a wide variation in the charters for PMOs
Project manager (PM)	The person held accountable by the project owner for the delivery of the project's outputs and for meeting the project plan	A project can have only one overarching project manager A PM can be contracted (from outside the funding organisation)
Project owner (PO)	The person held accountable by the funder(s) of the project for securing the project's target outcomes. The project owner acts on behalf of the funder(s) throughout the project—seeking to ensure that their interests are being served	Each project has only one project owner One entity can fill the roles of both funder and owner The project owner is the project manager's "client" A project owner must be appointed from within the funding organisation
Project plan	A baseline document that provides all the information required to: • Make a reliable decision about approving a start to work on a project • Track a project's progress	The project plan becomes the project's reference model—on which the management of execution is based The project-manager-designate is normally commissioned (by the project owner) to develop the project plan. The project manager is delegated responsibility for achievement of the project plan by the project owner
Project scope	A project is scoped when two conditions have been met: • All the outputs from the project have been identified and catalogued • Critical fit-for-purpose features have been set for each	
Project supplier	A "contracted" entity who provides products or services to the project manager	Suppliers can be internal or external
Project team	The group of individuals/organisations who produce the project's outputs	The project team is responsible for supporting the project manager in realising the project plan

(continued)

(continued)

Term	Definition/description	Discussion/examples
Project environment	That part of the business environment within which all four of a project's global phases are executed	Broadly, the business environment can be viewed as having two parts: • The project environment • The operational environment Refer also: business environment, operational environment
Project environment surveillance	The process used to manage external factors that are imposed on the project	Project environment surveillance is applied in particular to: Stakeholders, issues and risks
Project governance model (PGM)	A formal organisational model which: • Identifies all the stakeholders who will play a part in the project • Establishes the relationships and links amongst those stakeholders • Defines the roles of those stakeholders	A project governance model is separate from an organisational governance model—and, as a result, many of those involved in the project face a matrix management structure A PGM is linked into all the participating organisations through the line reporting arrangements of the individuals who will play a part in the project
Project management framework	A set of integrated, cohesive and related tools, procedures and techniques that can be used to guide the execution of a process	
Project worth	A notional index of a project's value. The "equation of worth" of a project involves three variables: • Benefits • Disbenefits • Costs	Worth represents a judgement by the funder about the net value of the project Benefits, disbenefits and costs (and hence worth itself) need not have a monetary value
Quality (output)	The quality of an output is defined by its list of fitness-for-purpose features ("FPFs")	
Rationale	A statement explaining why this particular project is being proposed at this particular time	
Realise (a target outcome)	To generate a target outcome	Outcomes are *realised*—while outputs are *delivered*
Reference group	A reference group is a formally constituted forum of key stakeholders—charged with filling a defined role in the project	Reference groups: • Report into the project (in the organisational sense) through either the project manager or the steering committee • Have highly specific terms of reference—and may also have a very short tenure
Register	A tabular catalogue of a particular class of project element in which instances are associated with rows and their attributes are shown in columns	The most common registers in a business case or project plan are: • Stakeholder • Issues • Risk

(continued)

(continued)

Term	Definition/description	Discussion/examples
Residual risk	The riskiness that remains after selected risk mitigation programmes have been adopted	
Resolving (an infeasibility)	Taking actions to reduce the level of project infeasibility. The available actions depend on both the type and severity of the infeasibility	All infeasibilities must be resolved before a project can be approved (or progressed) Refer also time-infeasible and cost-infeasible projects
Resource plan	A component of a project plan—indicating the full scope of resources that will be demanded by the project's work breakdown structure	A project has two resource plans: • An internal resource plan • An external resource plan The cost of a project is equal to value of the resources identified in both these plans If the project budget is less than the cost of the project, it becomes cost-infeasible
Resources	See inputs	
Return	Obtained as the worth of the project expressed as a % of total project cost	Analogous to the concept of return on investment (RoI) in investment theory
Risk	A risk is a single realisation of an event-impact model. This means that a risk has three components: • A threat (triggering event) • A chain of consequences • A final (damaging) impact	Refer also: attractiveness, riskiness
Risk management	An analytical/creative procedure for identifying, analysing, tracking and managing risks	
Risk manager	A member of the project team assigned responsibility for administering the risk management processes surrounding the project	
Risk Mitigation Programme (RMP)	A set of actions that will cause the overall expected damage (ED) of the project to fall	The actions that make up an RMP are of two kinds: • Pre-emptives—cause Li to fall • Contingencies—cause Se to fall An RMP requires additional resources
Risk register	A tool for cataloguing risks and documenting their analysis and management	Three participants in a project can enter threats into a risk register: the project owner, the project manager, the project assurance counsellor
Risk report	A regular report that highlights "important" risks for consideration by a project steering committee	A risk report is a subset (in terms of entries) of the risk register

(continued)

(continued)

Term	Definition/description	Discussion/examples
Riskiness	The level of uncertainty that the value set for project's worth in a business case (or project plan) will be realised	Obtained as the standard deviation of the spread of possible values of a project's worth Analogous to the concept of risk (as in the risk-return diagram) in investment theory
Role description	A specification of the role to be played in a project by an appointee to the PGM	Refer also: charter, terms of reference
Satisficing	A method of judging the success of a process using a criterion with an arbitrary value of acceptability	Term coined by Nobel prize Laureate Herbert Simon
Schedule	An array of data set out as a table—in which columns are associated with time periods	
Scope	A definitive statement about a project's boundaries	Principle: a project is scoped if and only if its outputs are defined
Scoping statement	A scoping statement for a project has five elements: • A statement of objective • Name of the entity filling the role of funder • A list of target outcomes • A list of committed outputs • A list of downstream processes	
Script	A formal description of how a process is executed	
Severity (Se)	A qualitative measure of the total reduction in the target worth of the project if a threat were to be realised	Severity is usefully gauged in terms of the same variable used to measure worth The Se of a threat before a Risk mitigation programme (RMP) is put into place is called "Pre-Se" The Se of a threat after an RMP is put into place is called "Post-Se"
Spontaneous stakeholding	A form of stakeholding arising when an entity's interest in a project arises regardless of any appointment to fill a role in the governance model	There are six generic classes of spontaneous stakeholder in a project: beneficiaries, customers, funders, impactees (positive), impactees (negative), influencers Refer also: commissioned stakeholding
Stage	One of a series of sequentially related sub-projects	The rationale for the overall project is to be found in the final stage Approval of a stage depends on the achievement of the target outcomes of the preceding stage

(continued)

(continued)

Term	Definition/description	Discussion/examples
Stakeholder	An individual or entity who is either: potentially impacted by the project; or who has a potential impact on the project	Anyone with "an interest in" the project is, by definition, a stakeholder in it Stakeholders are usefully separated into two classes: spontaneous and commissioned
Stakeholder engagement plan	A collection of activities that will be undertaken in order to appropriately engage a project's stakeholder	There are three generic forms of stakeholder engagement: • Include in the project communication plan • Make the subject of a special programme of engagement • Assign a role in the project governance model
Stakeholding	The stakeholding of a stakeholder in a project is defined by a list of the two-way impacts between the stakeholder and project	Refer also stakeholder
Statement of objective	A formal statement about the purpose of the project	
Steel tetrahedron	A suggested alternative (to the "iron triangle") test of success for project management	The "steel tetrahedron" requires that all the project's outputs be delivered: • To an agreed specification • On time • Within budget • Without detrimental outcomes attributable to the actions of the project manager The steel tetrahedron allows for trade-offs amongst the four criteria and thus establishes necessary and sufficient conditions for project management success Refer also: iron triangle
Steering committee (SC)	The body that is formally charged with supporting the project owner in discharging his/her accountabilities	The steering committee is made up of a small group of powerful supporters of the project
Success	A state of acceptable performance	Success is *judged* (or determined). By way of contrast, performance is *measured* (or gauged)
Tangibility	A characteristic of an artefact (or natural object) whereby it can be touched	Outputs, as artefacts, are always tangible. Outcomes as end-effects are never tangible (but they are measurable)
Target outcomes	A desirable outcome that is sought by the funder—and for which a threshold of achievement is set	Every target outcome from a project is represented by one or more downstream process metrics

(continued)

(continued)

Term	Definition/description	Discussion/examples
Task	A primitive process representing the lowest level of work recognised in a Work Breakdown Structure Also used loosely to mean any "block of work"	In the stylised three-level PAT (phase, activity, task) Work Breakdown Structure, processes at the third (lowest) level are called tasks. A collection of related tasks constitutes an activity Word structure: Tasks are expressed as imperatives (command phrases) Refer also: phase, activity
Team member	An entity that provides labour or produces outputs for the project	Can take either of two forms: • An individual or group who is assigned to the team to provide labour-based inputs to the project • An internal or external entity that is "subcontracted" to produce a specific project output
Terms of reference (ToR)	A specification of the role to be played in a project by an element of the project governance model—other than internal team members	A framework to guide a role defined in a projects' governance model. Also identified variously as: charter, brief, role description, job description Refer also: charter, role description
Threat	An event that has two characteristics: • It may or may not happen • If it does happen, it will eventually lead to a decrease in the worth of a project	Also called a "triggering event"
Timeframe	The overall duration of a project	There are two views of a project's duration: • The project manager's—measured from approval through to delivery of all outputs • The funder's/owner's—measured from approval through to realisation of target outcomes
Time-infeasible (project)	A project for which an optimal timeframe implied by its workplan exceeds the timeframe agreed-to by the owner/funder	In such a situation, deadlines fall earlier than the milestones derived from an optimal workplan. The difference between a deadline and the corresponding milestone is a gauge of the level of infeasibility of the project
Underengineering (of an output)	A situation in which the definition of an output is missing critical fitness-for-purpose features)	If an output is underscoped, it cannot be utilised effectively by a project customer —and hence the generation of target outcomes may be compromised

(continued)

(continued)

Term	Definition/description	Discussion/examples
Underscoping (of a project)	A situation in which the current scope of a project is missing critical outputs (or some outputs are missing necessary fitness-for-purpose features)	If a project is underscoped, it cannot achieve its target outcomes
Undesirable outcomes	An outcome that reduces the worth of the project in the eyes of the project funder	Whereas a desirable outcome can only emerge from a project after its outputs have been delivered, an undesirable outcome can emerge from a project before outputs have been produced
User	A term from IS/IT that is not used in the ITO model	A corresponding defined term in the ITO model is project customer
Utilisation	The employment of an output by a project's customer(s) that can cause target outcomes to emerge as a by-product	
Word structure	The permissible ways in which particular instances of terms from this glossary are to be expressed	(This is a meta-term within this glossary—it is used to construct the glossary—but is not, itself, a project management term)
Work breakdown structure (WBS)	A hierarchical model (structured breakdown) of the work involved in executing a project	The term "WBS" can be applied to the work involved in producing a single project output—or to the project as a whole Refer also: phase and task
Work package	An element of the Work breakdown structure that can be assigned to a team member and later tracked as a single "block of work"	Any of the three levels in a PAT-style WBS may qualify as a work package in particular situations

Appendix B
Project Governance: Role Definitions

This Appendix provides a detailed description of the nine core project roles discussed in Sect. 4.1.6. Some of the text appearing in the "Description" columns of the following tables is *italicised, used* for guidance when creating a new entry for a specific application of the role definition. It would be replaced with text relevant to the particular role being defined. All non-italicised text would appear basically as it stands.

B.1 Steering Committee

Role:
The steering committee (SC) supports the project owner in looking after the interests of the funder. It seeks to ensure that the business case and project plan will be realised, and that project is always "pointed at" its baseline documents.

Principles:

- There can be only one SC, made up of a small group of powerful supporters of the project. An implication of this principle is that anyone who opposes the project is automatically disqualified from membership (If an opponent of the project is to play a role at all, it would be as a member of a reference group).
- The SC works almost exclusively above the line and is not an alternative or (worse still) a *competing* project team. Consider a project where disagreement has broken out amongst team members over the design of a high-quality (and expensive) brochure. It would be appropriate to identify this as an issue for consideration by the SC (from whom guidance might be sought on how the matter is to be resolved), but it would not be appropriate to have the SC make the design decision. In the case of projects with multiple funders, each may (in consultation with the others) appoint a representative to the SC.

O. Zwikael and J. R. Smyrk, *Project Management*,
https://doi.org/10.1007/978-3-030-03174-9

- The project owner tables proposals for changes to baseline documents and seeks the SC's approval for those changes.
- The project owner is normally the chair of the SC.
- A succession of short-term appointments is highly undesirable because it can lead to an SC degenerating into a steering *parade*. It is reasonable to expect that projects will be no more stable than their SCs.

Membership:

- A funder can be a member of an SC.
- The role of the SC is such that no one should be appointed as a member in order to gain access to his/her technical expertise. That sort of involvement would be best accommodated by making the participant an advisor or a member of a reference group (or even a member of the project team).
- A trade-off has to be made with SC size. The larger the committee, the slower and more cumbersome the decision cycle. If, on the other hand, key players are missing, then the ability of the committee to positively influence the course of the project may be compromised.
- Because the SC will, in consultation with the project owner, approve changes to the project's baseline documents, membership is restricted by consideration of conflicts of interest. Not only are external suppliers clearly disqualified from membership, but there could well be cases where such a restriction might also apply to internal suppliers as well.

Qualifications:
An SC member should be strongly committed to the success of the project. In addition, he/she should be able to add value to the work of the committee by being politically influential, a reliable decision-maker and/or a strong leader. Membership of an SC is onerous and not to be taken lightly. In general, the role is poorly understood amongst the executive management community with the result that many SC meetings become little more than exercises in communication rather than forums for decision-making. It is highly desirable to have members of SCs exposed to some form of formal induction programme (even if only short) to ensure that members understand the role and are able to participate meaningfully.

Engagement:
A charter is established for the SC (based on a generic structure that would apply to most projects). The project funder would invite candidates to join (or appoint them). Each candidate is required to formally accept the SC charter. Assembly of the SC would normally be done during project planning under the leadership of the project owner with administrative support from the project manager.

Table B.1 A generic role definition/template for the steering committee

Attribute	Description
Name of PGM role	Steering committee (SC)
Objective of the role	The steering committee seeks to ensure that the project and its baseline documents are always aligned
Outputs from role	The SC issues guidance and instructions to the project manager and to any reference group which reports directly to it
Core activities and frequencies	The SC conducts regular project review meetings Members undertake tasks assigned during those meetings
Membership	Members of the SC are (*Here list all members*) The chair of the SC is (*name of chair* (*typically the PO*))
Communications programme	Communication with the SC will be via formal periodic project status reports (*Where additional communication appears appropriate, it should be identified (but not defined) as another entry in this cell*)
To whom does this role report	The SC does not report directly outside the PGM, however, each member reports to the funding entity via the positions members hold in the organisation's standing structure
Review of role (who/when)	The funder will review and assess the performance of the SC (*nominally every 3 months*)
Dates of creation and termination	The SC will be established on <date> (*this could be as early as the beginning of planning but should be no later than the beginning of execution*) and disband on <date> (*this would normally be set as the date when the achievement of target outcomes is expected*)

Performance evaluation:
A SC should be assessed collectively by the funder(s) after a project's target outcomes are secured. That assessment should be based on the extent to which the last baseline business case was realised.

Support:
The project manager and/or project administrator serves as the secretariat for the SC.

A generic role definition:
Table B.1 provides a generic structure for the role of the SC.

B.2 Project Owner

Role:
As the funder's agent, the project owner (PO) is fully accountable to the project's funder for realising the business case in general and securing the flow of target outcomes in particular. Ideally, the project owner will chair the steering committee, but in some cases, there may be better-qualified members to fill that role.

Principles:

- The PO would normally be appointed from inside the funding entity.
- There can be only one PO.
- From a pseudo-commercial viewpoint, the PO is the project manager's client.

Qualifications:
A PO should be strongly committed to the project and have a basic understanding of project planning and management. A PO must be capable of dealing with the numerous political issues that arise in the course of a project.

Engagement:
A role definition is established for the PO (based on a generic structure that would apply to most projects). He/she is engaged by the funder. The project champion is, in many cases, an obvious candidate for this role. The PO is required to formally accept the engagement. This would normally be done by the time that a decision is taken in favour of the business case.

Performance evaluation:
The PO should be assessed by the funder after a project's target outcomes are secured. That assessment (as in the case of the SC) should be based on the extent to which the last baseline business case was realised.

A generic role definition:
Table B.2 provides a generic structure for the role of the PO.

Table B.2 A generic role definition/template for the project owner

Attribute	Description
Name of PGM role	Project owner (PO)
Objective of the role	The project owner serves as • The funder's agent—ensuring that the project is conducted in the funder's best interests • The project manager's purchaser/client The PO • Maintains close ongoing contact with: the funder, the PM, members of the SC, the project assurance counsellor and the probity counsellor • Participates in SC meetings (ideally as chair)
Accountability	The PO is accountable to the funder for the achievement of the business case in general and for the realisation of target outcomes in particular
Outputs from role	The PO issues guidance and instructions to the project manager
Core activities and frequencies	The PO • Maintains close ongoing contact with: the funder, the PM, members of the SC, all reporting reference groups, the project assurance counsellor and the probity counsellor • Participates in SC meetings (ideally as chair)

(continued)

Table B.2 (continued)

Attribute	Description
To whom does this role report	The PO reports to the funder
Review of role (who/when)	The funder will review and assess the performance of the PO (nominally every three months)
Dates of creation and termination	The PO will be appointed on <date> (*this could be as early as the beginning of initiation but should be no later than the beginning of planning*) and step down on <date> (*this would normally be set as the date when the achievement of target outcomes is expected*)

B.3 Project Manager

Role:
The project manager (PM) is accountable to the PO for delivering the project's committed outputs, fit-for-purpose, within the constraints of an agreed budget and timeframe and without causing undesirable outcomes.

Principles:

- PMs can come from inside or outside the funding entity.
- The PM is engaged by the PO in consultation with the funder.
- There can be only one PM; however, under him/her there may be multiple sub-PM or team leaders, each one responsible for particular outputs and/or certain areas of work. In many such cases, the PM may be identified as a programme manager and those reporting as "project managers". If this principle is violated (resulting in more than one PM) then the question arises who is accountable for resolving conflicts about output delivery. The answer, by definition, identifies a PM (and so logically there can be only one).
- By implication, a PM is accountable only to the steering committee (through the PO). The PM must not be required to report (in the line sense) to any other entity in the project governance model, especially reference groups.
- The PM would be expected to chair a project control group in those cases where it is required.
- The PM (or appropriate delegates) serves at the SC secretariat.

Qualifications:
A PM should be strongly committed to the project and have thorough knowledge of project management methodologies. The larger the project, the more desirable it is that the PM has professionally recognised credentials and knowledge of effective analysis techniques (e.g. Zwikael et al. 2015). Like the PO, a PM must be politically astute, capable of managing upwards in his/her dealings with the steering committee, reference groups and advisers and of critical thinking (Frank et al. 2007).

Table B.3 A generic role definition/template for the project manager

Attribute	Description
Name of PGM role	Project manager (PM)
Objective of the role	The project manager serves as the project owner's provider/supplier (of outputs)
Outputs from role	The PM issues guidance and instructions to the project team and any reference group that reports directly to him
Core activities and frequencies	The PM • Maintains close ongoing contact with: the PO, members of the PCG, all reporting reference groups, the project assurance counsellor and the probity counsellor • Participates in PCG/team meetings (as chair)
To whom does this role report	The PM reports to the PO
Review of role (who/when)	The PO (in consultation with the SC) will review and assess the performance of the PM at agreed intervals
Dates of creation and termination	The PM will be appointed on <date> (*this could be as early as the beginning of initiation, but should be no later than the beginning of planning*) and step down when all outputs have been delivered

Engagement:
A role definition for the PM is based on a generic structure that would apply to all projects. The person who assisted the champion to assemble the business case is obviously a strong candidate to fill the role of PM. This would normally be done as soon as a decision is taken in favour of the business case.

Performance evaluation:
The PM should be assessed periodically throughout project execution by the PO and again after a project's outputs are delivered. That assessment should be based on the "steel" tetrahedron (outlined in Sect. 3.2) involving criteria for: scope/quality, timeframe, cost and undesirable outcomes.

A generic role definition:
Table B.3 provides a generic structure for the role of the PM.

B.4 Project Administrator

Role:
The project administrator (PA) is accountable to the PM for guiding, coordinating and monitoring all Above-the-Line work.

Principles:

- PAs can come from inside or outside the funding entity.
- The PA is engaged by the PM in consultation with the PO.
- The PA is the PM's agent, while the project manager is the PA's principal.

Table B.4 A generic role definition/template for the project administrator

Attribute	Description
Name of PGM role	Project administrator (PA)
Objective of the role	The project administrator seeks to ensure that all Above-the-Line processes are executed successfully
Outputs from role	The PA issues guidance and instructions on the conduct of Above-The-Line processes to all participants in the project governance model
Core activities and frequencies	The PA • Maintains close ongoing contact with all participants in the project governance model • Participates in PCG/team meetings (as chair)
To whom does this role report	The PA reports to the PM
Review of role (who/when)	The PM (in consultation with the PO) will review and assess the performance of the PAA at agreed intervals
Dates of creation and termination	The PA will be appointed on <date> (*this could be as early as the beginning of initiation but should be no later than the beginning of planning*) and step down when all outputs have been delivered

Qualifications:
A PA should have thorough knowledge of project management methodologies. The larger the project, the more desirable it is that the PA has professionally recognised credentials. A PA must have well-developed interpersonal skills.

Engagement:
A role definition for the PA is based on a generic structure that would apply to all projects in the organisation.

Performance evaluation:
The PA should be assessed periodically throughout project execution by the PM and again after a project's outputs are delivered. That assessment should be based on the performance of the project's Above-the Line processes.

A generic role definition:
Table B.4 provides a generic structure for the role of the PA.

B.5 Project Control Group

In large exercises, the PM's span of control may involve a considerable number of direct reports—thus creating issues for the effectiveness of his/her management. This may be addressed by creating a project control group (PCG). The PCG is made up of all those who manage significant pools of resources such as team leaders and representatives of suppliers, subcontractors or consultants. A PCG is similar in *structure* to the steering committee (SC)—but whose members have accountabilities delegated by the PM. The PCG does not interact directly with the SC—this remains as part of the PM's role. The PCG is somewhat similar to the concept of an advisory board or project board.

Role:
A PCG supports the PM during the global phase of project execution.

Principles:

- The PCG works both above-the-line and below-the-line.
- A PCG member may fill a "representation" role on behalf of the organisational entity that he/she represents.
- The PM is normally the chair of the PCG.

Membership:

- The PCG is made up of a small group of senior influential members of the project team. Members would typically be drawn from: sub-teams, suppliers, subcontractors and consultants.
- As is the case with the SC, the size of the PGC involves a trade-off between the speed of the decision cycle and the scope of influence it can exert over project execution.

Qualifications:
A PCG member should be able to exercise significant influence over the way the area he/she represents participates in project execution. In addition, he/she should be strongly committed to the project, a reliable decision-maker and/or a strong leader.

Engagement:
A charter is established for the PCG (based on a generic structure which would apply to most projects). The PM (in consultation with the PO) would invite candidates to join (or appoint them). Each candidate is required to formally accept the PCG charter. Assembly of the PCG would normally be done during project planning under the leadership of the PO with administrative support from the PM.

Performance evaluation:
As in the case of the PM, the PCG should be assessed periodically throughout project execution by the PO and again after a project's outputs are delivered. That assessment is also based on the "steel" tetrahedron.

Support:
The PM and/or project administrator serves as the secretariat for the PCG.

A generic role definition:
Table B.5 provides a generic structure for the role of the PCG.

Table B.5 A generic role definition/template for the project control group

Attribute	Description
Name of PGM role	Project control group (PCG)
Objective of the role	The project control group supports the PM discharging his/her accountabilities
Outputs from role	The PCG issues guidance and instructions to the project team—particularly suppliers, consultants, contractors
Core activities and frequencies	The PCG conducts regular project review meetings Members undertake tasks assigned during those meetings
Membership	Members of the PCG are (*Here list all members*) The chair of the PCG is the PM
To whom does this role report	The PCG reports to the PM
Review of role (who/when)	The PO and PM will review and assess the performance of the PCG (*nominally every three months*)
Dates of creation and termination	The PCG will be established on <date> (*this would normally be early in execution*) and step down when all outputs have been delivered

B.6 Project Team

Role:
The project team builds, assembles, delivers and implements the project's committed outputs. Project teams often demand a formal structure of their own, especially if they are made up of sub-teams and suppliers. Appropriate structures may be based on outputs, skills or even the affiliations of team members.

Members of the project team are accountable to a designated leader/manager for completing assigned tasks (or producing assigned outputs). Project team members may also be assigned specific above-the-line responsibilities (such as managing the risk register). In larger projects, the demands of above-the-line work become so great that it becomes necessary to support the team with specialist project administrators.

Qualifications:
Project team members will be appointed according to their capabilities to undertake selected parts of the project's WBS. Project administrators must have a deep knowledge of project management. It is also expected that they would be persistent and relentless with detail.

Engagement:
A role definition is established for each project team member. He/she is engaged by the PM, or, if the project is large, by an appropriate sub-team leader or contractor. In many projects, key team members are appointed from the ranks of the participating organisations. A number of issues can arise with internal appointments that

Table B.6 A generic role definition/template for a project team member

Attribute	Description
Name of role	(*A "label" which uniquely identifies this particular member role*)
Objective of the role	(*Why is this role being established?*)
Outputs from role	(*What outputs will flow from the work of the role?*)
Core activities and frequencies	The <name of team member role> (*Here provide a simple list the most important activities in which this team member will be involved—and an indication of the frequency with which those activities will be undertaken*)
To whom does this role report	(*Identify where this role reports. In small projects, it will be the PM, but in larger exercises it will be the next immediate level up the PGM—such as a team leader*)
Review of role (who/when)	(*Who is to review the performance of this role and when is this to happen?*)
Dates of creation and termination	The <name of team member role> will be appointed on <date> and conclude on <date> (*These dates may be replaced with events*)

may require resolution. Two are worthy of particular note. Because staff are effectively "offline" while working on projects, their career development may suffer, especially if the appointment is long-term and full-time. Part-time appointments and short-term secondments bring with them issues of resource contention (because staff now face dual reporting lines). In those cases, it may be desirable to cover the assignment with a memorandum of understanding between the staff member, the relevant line manager and the PM (see Sect. 4.3.2). Such a document would openly summarise the impacts of the appointment on the staff member and map out a plan to address them, as well as establishing ground rules for resource allocation between competing demands.

Performance evaluation:
Each team member should be assessed by the manager to whom they report in the team structure using a "local" variant of the steel tetrahedron. Evaluations of those staff drawn from the participating organisations should be provided to HR for incorporation into their personnel records.

A generic role definition:
Table B.6 provides a generic structure for the role of a project team member.

B.7 Reference Groups and Advisers

Role:
Reference groups provide specialist input to the project. Reference groups (or "advisers" if a single person is involved) are highly specific and so it is not uncommon to find a large number of these represented in the model (although not all of these may exist at the one time). Reference groups, as well as advisers, support the team in a similar way to subcontractors.

Principles:
The decision to locate someone to the reference (rather than the delivery) division of the model is, in general, based on a number of criteria:

- The first concerns the way the role reports into the governance model. If a role is to report to the steering committee, then it cannot be located in the delivery division. A lawyer commissioned to review the PM's contract, for example, would be recognised as an adviser to the SC.
- The second concerns the nature of the resources provided. If someone is not a "deployable" resource, they will probably be recognised as an adviser. For example, if a senior manager from IT is commissioned to confirm the acceptability of a software product selected by the team, then he/she is better viewed as an adviser (Zwikael et al. 2008).
- The third concerns the intensity and duration of involvement. It may not be convenient to have someone who has a relatively minor, periodic involvement with the project assigned to a formal position in the project team structure.
- The fourth relates to the need for independence. Someone who is to comment on (or validate) the work of the team may be better located outside the team.

In some cases, it is an arbitrary decision whether these are located in the reference or the delivery division of the governance model.

Qualifications:
Reference group members and advisers will be appointed according to their capabilities to undertake selected parts of the project's WBS.

Engagement:
A role definition is established for each reference group member and adviser. Depending on the brief, he/she is engaged by either the PM or SC. Reference group members and advisers can come from inside or outside the performing organisation. The issue of remuneration for reference groups and advisers will be resolved according to the circumstances surrounding each proposed engagement.

Performance evaluation:
Where appropriate, the evaluation of reference groups and advisers would normally be undertaken as part of their line roles within their organisation's standing structure.

A generic role definition:
Table B.7 provides a generic structure for the role of the reference groups and advisers.

Table B.7 A generic role definition/template for reference groups/advisers

Attribute	Description
Name of role	(*A "label" which uniquely identifies this* reference group/adviser *role*)
Objective of the role	(*Why is this role being established?*)
Outputs from role	(*What outputs will flow from the work of the role?*)
Core activities and frequencies	(*Here provide a simple list the most important activities in which this reference group/adviser will be involved—and an indication of the frequency with which those activities will be undertaken*)
To whom does this role report	(*Identify where this role reports. In small projects, it will be the either the PO/SC or PM/PCG, but in larger exercises it may be to sub-project CGs /PMs*)
Review of role (who/when)	(*Who is to review the performance of this role and when is this to happen?*)
Dates of creation and termination	The <name of reference group/adviser role> will be appointed on <date> and conclude on <date> (*These dates may be replaced with events*)

B.8 Project Assurance Counsellors

Role:
The project assurance counsellor (PAC) ensures that the project is being run in accordance with accepted project management practice. Projects need not have a project assurance counsellor, but the larger the exercise, the stronger the case that can be made for appointing one. In smaller projects, a Project Management Office may be qualified to fill this role.

Project assurance counsellors are expected to:

- Monitor the way that the project's above-the-line work is being carried out.
- Work with the PM to resolve issues in this area.
- Report periodically to the steering committee.
- Project assurance is an advisory rather than an audit role.

Principles:
A project assurance counsellor must not face any conflicts of interest in his/her appointment.

Qualifications:
Project assurance counsellors are required to have a deep understanding of project management or commercial practice. They would also normally be expected to have well-developed political and organisational skills.

Engagement:
Project assurance counsellors can come from inside or outside the participating organisation and are appointed by the SC (in consultation with the PO and PM).

Table B.8 A generic role definition/template for project assurance a counsellor

Attribute	Description
Name of role	Project assurance counsellor (PAC)
Objective of the role	The PAC is to ensure that the project is being conducted in accordance with whatever framework of management has been adopted for the exercise
Outputs from role	Primary outputs from the PAC role include • Regular briefings and reports to the SC • Ad hoc advice and guidance to the PO and PM
Core activities and frequencies	• Ad hoc and periodic formal interaction with the PO • Ad hoc and periodic formal interaction with the PM • Preparation of regular project assurance reports • Periodic presentations to the SC
To whom does this role report	The PAC reports to the SC
Review of role (who/when)	The PO will review the role of the PAC at agreed points throughout the project
Dates of creation and termination	The PAC will be appointed on <date> (*this could be as early as the beginning of initiation but should be no later than the beginning of planning*) and step down on <date> (*this would normally be set as the date when the achievement of target outcomes is expected*)

Performance evaluation:
The evaluation of a project assurance counsellor would normally be undertaken periodically by the SC.

A generic role definition:
Table B.8 provides a generic structure for the role of the project assurance counsellors.

B.9 Probity Counsellor

Role:
A probity counsellor ensures that the commercial dealings between the project team and the outside world are consistent with the policies, standards, regulations, laws and practices that apply to the project environment. Projects need not have a probity counsellor, but the more significant the outlays for external resources, the stronger the case that can be made for appointing one.

Probity counsellors are expected to:

- Monitor the entire procurement process.
- Work with the PM to resolve issues in this area.
- Report periodically to the steering committee.

As is the case with project assurance counsellors, this is an advisory rather than an audit role. Any audits should be assigned to either the audit function within the funding organisation, or an external firm of professionals (Such an appointment would then be reflected in the addition of an appropriate advisory function in the Project governance model).

Principles:
A probity counsellor must not face any conflicts of interest in his/her appointment.

Qualifications:
Probity counsellors are required to have a deep understanding of local project procurement practice.

Engagement:
Probity counsellors can come from inside or outside the participating organisation and are appointed by the SC (in consultation with the PO and PM).

Performance evaluation:
The evaluation of a probity counsellor would normally be undertaken periodically by the SC.

A generic role definition:
Table B.9 provides a generic structure for the role of the probity counsellor.

Table B.9 A generic role definition/template for probity counsellor

Attribute	Description
Name of role	Probity counsellor
Objective of the role	The probity counsellor is to ensure that the commercial dealings between all project participants and the outside world accord with the standards, conventions, rules, regulations set by the funder (or imposed by the project's institutional/statutory context)
Outputs from role	Primary outputs from the PAC role include • Regular briefings and reports to the SC • Ad hoc advice and guidance to the PO and PM
Core activities and frequencies	• Ad hoc and periodic formal interaction with the PO • Ad hoc and periodic formal interaction with the PM • Preparation of regular project probity reports • Periodic presentations to the SC
To whom does this role report	The probity counsellor reports to the SC
Review of role (who/when)	The funder will review the role of the probity counsellor at agreed points throughout the project
Dates of creation and termination	The PAC will be appointed on <date> (*this could be as early as the beginning of initiation but should be no later than the beginning of planning*) and step down on <date> (*this would normally be set as the date when the achievement of target outcomes is expected*)

Appendix C
Questions for Future Research

Abstract

The book discusses a comprehensive methodology for the management of projects to enhance the eventual flow of desired benefits. Because project benefit management is, however, still a relatively immature research area, there are many potential ideas that other researchers can explore. Therefore, this appendix suggests areas of future research for students, scholars and practitioners to continue developing knowledge and tools, building on the concepts discussed in the book. These research questions are grouped according to major project management topics.

C.1 Research Questions on Project Success

1. The book proposes a general project success model consists of three dimensions (project management success, project ownership success and project investment success).
   - How can established dimensions of project success be mapped into the proposed model?
   - What scales and items can be developed to measure the three dimensions of the success model?
   - What are the success rates in practice for each of the three dimensions of the success model? Do they differ among countries, economies, industries, organisations and project types? If so how?
2. The model uses the concept of a regression test—based on the project's baseline documents (i.e. the business case and project plan). The initial versions may, however, be progressively updated as execution progresses.

O. Zwikael and J. R. Smyrk, *Project Management*,
https://doi.org/10.1007/978-3-030-03174-9

- Which version of the project's baseline documents should be used in the application of the regression test (and hence when making judgements about each of the three forms of project success)?

C.2 Research Questions on Project Outcomes

1. Section 2.4.2 suggests a taxonomy of project outcomes.
   - Do outcomes have an underlying generic taxonomy—or do taxonomies vary according to contextual parameters—such as country, economic sector, industry organisations and project type? How might a useful taxonomy be structured?
   - How might the usefulness/effectiveness of such a taxonomy be gauged?
2. The ITO model postulates that, in general, the generation of target outcomes is "caused by" the utilisation of committed outputs.
   - What is the nature of the relationship between producing project outputs and realising project outcomes?
   - If there is weak causality between outputs and outcomes, can target outfcomes be achieved without the production of the project's committed outputs? Under what conditions?
3. Section 2.5.4 discusses "Natural" outcomes—which appear automatically—but only when an (existing) operational process is terminated (i.e. it is not necessary that a committed output be utilised by a project customer to realise this outcome).
   - Is it possible to observe natural outcomes under any other conditions—for example, when a committed output *is* utilised by a project customer?
   - Is there a meaningful target outcome for a project that aims to achieve "compliance"? For example, compliance with a new government regulation to undertake annual safety audits. Or is "compliance" merely *the state of having fulfilled some condition* (such as the introduction of the required audits)?
4. Section 2.4.6 distinguishes between "relative" and "absolute" labels for target outcomes. This is based on the acceptance of a distinction in the 2NY Map between "Now" values *which don't exist* and "Now" values *of zero*.
   - Is it meaningful to claim that a variable which has never been observed is the same things as variable which has been observed to have a value of zero?
   - What empirical evidence exists in research for the impact of effective target outcomes on enhanced organisational performance in practice?
   - How many target outcomes should a project have? What are the implications of setting "too many" and "too few"?
   - Could the concept of discounting also be applied to the project's non-monetary outcomes? How?

C.3 Research Questions on Operations Management

1. Figure 2.2 proposes a trichotomy of work: ad hoc tasks, operational processes and projects.
   - Is this a true trichotomy, or do the three classifications lie on a continuum of some kind?
   - What is the difference between a "complex" operational process and a simple project? Is there an overlap between the two? What are the implications of assigning a measure of complexity?
   - Under what conditions is an operations manager a suitable candidate for the project owner role in operations enhancement projects?
2. Section 9.3.4 introduces the outcomes coverage matrix (OCM). The OCM requires that a judgement be made about whether a target outcome can be used as a performance metric for a downstream process—whereby, in the OCM. "1" means "yes", while "0" means "no".
   - Should the entries in the outcomes coverage matrix (OCM) indicate *the amount of change* in the metric of an operations process that the project can bring about?
3. Consider two projects A and B with totally independent OCMs. That is, any "1"s in the OCM for A relate target outcomes and downstream processes which have only "0"s in the OCM for B. If this independence is not recognised, a single OCM will be adopted for the combined project C (=A + B).
   - How can the OCM for C be used to partition it into its two independent components?

C.4 Research Questions on Project Governance

1. Section 3.2 claims that the effective management of a project's key players requires an overarching governance model.
   - What is the underlying nature of the relationship between effective project governance and outcome realisation?
   - What project governance practices are most effective in improving project success? How could "effectiveness" be measured empirically?
   - Does the assignment of an accountability for outcome realisation improve project success rates?
   - What are the most effective management approaches, practices and tools to support the role of the project owner?
   - What are the characteristics of an effective project owner?

- Principal–agent relationship exists between project funder and project owner. Does principal–agent relationship also exist between project owner and project manager?
- In what type of projects is the role of the project owner most effective (e.g. simple vs complex projects, clear target outcomes vs undefined goals)?
- During execution, it may prove necessary to vary certain parameters that were set in the accepted business case. If these changes lie outside certain boundaries, the project owner will need the approval of the funder. How should these boundaries be established?
- What are effective design guidelines for the structure of the delivery division?

C.5 Research Questions on Project Scoping

1. Each output has a set of fitness-for-purpose feature.
    - Under what conditions will a fitness-for-purpose feature (FPF) evolve into a new output?
    - Under what conditions should a collection of related outputs be treated as one?

C.6 Research Questions on Project Planning

1. In Chap. 9 we discuss the "rule of 7-ish"—which suggests that each element of a WBS should be decomposed into "about seven" components. As the count of components increases, the "depth" (count of nodes) of the hierarchy decreases, but the count of elements in each node increases.
    - How should the effectiveness of a hierarchical decomposition count be assessed?
    - What is an optimal hierarchical decomposition count?
2. In Chap. 10 we suggest a minimal level of involvement by the funder in planning.
    - What sorts of involvement could be classified as appropriate/inappropriate?
    - What are the determinants of appropriate forms of involvement by a funder in project planning?

C.7 Research Questions on Risk

1. The event-impact model of risk (Sect. 6.2.3) assumes event-independence—whereby the realisation of one threat has no effect on the likelihood or severity of another.
   - How could risk interdependencies be integrated into the analysis of threats, project riskiness and incorporated into the attractiveness map?
2. Utility theory can be used to characterise attitudes to risk. For example, a risk-averse (risk-avoiding) individual values the result of an uncertain event at less than its expected value. The reverse is true for a risk-affine (risk-seeking) individual.
   - How might utility theory be integrated into the analysis of project riskiness and the attractiveness map?
3. The approach to risk adopted in the book is *asymmetric* (refer to Sect. 6.2.2), because it recognises only downside risk.
   - How could opportunities (together with threats) be integrated into a risk management model (and hence into the risk register)?
   - What would a symmetric approach to risk imply for project risk management?
   - What implications for risk management would arise from use of a *symmetric* model of risk which in addition to threats also takes into account opportunities?
4. Some generic threats appear regularly in the risk registers across all projects—for example, "Project manager resigns", while others appear to be particular to certain types of projects (such as "bad weather during construction").
   - Is there a useful underlying taxonomy that might be used to classify generic threats? How could the classes in such a taxonomy be defined?
   - Should threat severity be measured in dollar terms or as a percentage of the total project worth?

Appendix D
Reference List

Aubry, M. (2015). Project management office transformations: Direct and moderating effects that enhance performance and maturity. *Project Management Journal, 46*(5), 19–45.

Baker, B. N., Murphy, D. C., & Fisher, D. (1988). Factors affecting project success. In D. I. Cleland, & W. R. King (Eds.), *Project management handbook* (pp. 902–919). Van Nostrand Reinhold: New York.

Bourne, L., & Walker, D. H. T. (2005). Visualising and mapping stakeholder influence. *Management Decision, 43*(5/6), 649–660.

Campbell, H., & Brown, R. (2003). Benefit–Cost analysis. *Financial & economic appraisal using spreadsheets*, Cambridge.

Chih, Y., & Zwikael, O. (2015). Project benefit management: A conceptual framework of target benefit formulation. *International Journal of Project Management, 33*(1), 352–362.

Cicmil, S., Williams, T., Thomas, J., & Hodgson, D. (2006). Rethinking project management: Researching the actuality of projects. *International Journal of Project Management, 24*(8), 675–686.

Crawford, L., Pollack J., & England D. (2006). Uncovering the trends in project management: Journal emphases over the last 10 years. *International Journal of Project Management, 24*(2), 175–184.

Daft, R. L. (2007). *Understanding the theory and design of organizations.* Thomson: South-Western.

Dai, C. X., & Wells, W. G. (2004). An exploration of project management office features and their relationship to project performance. *International Journal of Project Management, 22*, 523–532.

Daniel, R. H. (1961). Management data crisis. *Harvard Business Review. Sept–Oct*, 111–112.

Donaldson, L. (2001). The contingency theory of organizations. In *Thousands Oaks*, CA: Sage.

Dunar, A. J., & Waring, S. P. (1999). Power to explore—history of marshall space flight center 1960–1990. U.S. Government Printing Office. ISBN 0-16-058992-4.

Dvir, D., & Lechler, T. (2004). Plans are nothing, changing plans is everything: The impact of changes on project success. *Research Policy, 33*, 1–15.

Eisenhardt, K. M. (1989). Agency theory: An assessment and review. *The Academy of Management Review, 14*(1), 57–74.

El-Gohary, N. M., Osman H., & El-Diraby T. E. (2006). Stakeholder management for public private partnerships. *International Journal of Project Management, 24*(7), 595–604.

Flyvbjerg, B. (2005). Design by deception: The politics of megaproject approval. *Harvard Design Magazine, Spring/Summer, 22*, 50–59.

O. Zwikael and J. R. Smyrk, *Project Management*,
https://doi.org/10.1007/978-3-030-03174-9

Flyvbjerg, B. (2017). The iron law of megaproject management. *The Oxford handbook of megaproject management*, Oxford University Press.
Flyvbjerg, B., Ansara, A., Budzier, A., Buhl, S., Cantarelli, C., Garbuio, M., Glenting, C., Holm, M.S., Lovallo, D., Lunn, D., Molin, E., Rønnest, A., Stewart, A., & van Wee, B. (2018). Five things you should know about cost overrun. *Transportation Research Part A, 118*, 174–190.
Ford, J. D., Ford, L. W., & D'Amelio, A. (2008). Resistance to change: the rest of the story. *The Academy of Management Review, 33*(2), 362–377.
Fortune, J., & White, D. (2006). Framing of project critical success factors by a systems model. *International Journal of Project Management, 24*(1), 53–65.
Frank, M., Zwikael, O., & Boasson, M. (2007). Jobs requiring a capacity for engineering systems thinking (CEST): Selection using an interest inventory. *Project Management Journal, 38*(3), 36–44.
Gaddis, P. O. (1959). The project manager. *Harvard Business Review, 32*, 89–97.
GAO. (1994). *Advanced automation system: Implications of problems and recent changes.*
Globerson, S., & Zwikael, O. (2002). Impact of the project manager on project management planning processes. *Project Management Journal, 33*(3), 58–64.
Harvey, J., & Aubry, M. (2018). Project and processes: A convenient but simplistic dichotomy. *International Journal of Operations & Production Management, 38*(6), 1289–1311.
Hill, G. M. (2008). The complete project management office handbook (2nd ed.). Auerbach Publications, Taylor & Francis Group, Boca Raton, FL.
Hurt, M., & Thomas, J. (2009). Building value through sustainable project management offices. *Project Management Journal, 40*(1), 55–72.
Invernizzi, D.C., & Locatelli, G. (2018). What makes a (nuclear) decommissioning project successful? The Academy of Management Annual Meeting, Chicago, Il.
Jenner, S. (2009). Realising benefits from government ICT investment—A fool's errand? Academic Publishing, Reading.
Jensen, A., Thuesen, C., & Geraldi, J. (2016). The projectification of everything: Projects as a human condition. *Project Management Journal, 47*(3), 21–34.
Jha, K. N., & Iyer, K. C. (2007). Commitment, coordination, competence and the iron triangle. *International Journal of Project Management, 25*(5), 527–540.
Johnson, J., Karen, D., Boucher, K. C., & Robinson, J. (2001). The criteria for success. *Software Magazine, 21*(1), S3–S11.
Jugdev, K. (2004). Through the looking glass: examining theory development in project management with the resource-based view lens. *Project Management Journal, 35*(3), 15–26.
Kahneman, D. (2011). *Thinking, fast and slow* (1st ed.). New York: Farrar, Straus and Giroux.
Kellogg Foundation. (2004). *Logic model development guide.* Michigan: W.K. Kellogg Foundation.
Kerzner, H. (2013). *Project management: A systems approach to planning, scheduling and controlling* (11th ed.). Wiley.
Koh, A., & Crawford, L. (2012). Portfolio management: The Australian experience. *Project Management Journal, 43*(6), 33–42.
Korhonen, T., Laine, T., Lyly-Yrjänäinen, J., & Suomala, P. (2016). Innovation for multiproject management: The case of component commonality. *Project Management Journal, 47*(2), 130–143.
Lechler, T., & Dvir, D. (2010). An alternative taxonomy of project management structures: Linking project management structures and project success. *IEEE Transactions on Engineering Management, 57*(2), 198–210.
Letavec, C. J. (2006). The program management office: establishing, managing and growing the value of a PMO, J. Ross Publishing, Fort Lauderdale, FL.
Lipke, W., Zwikael, O., Henderson, K., & Anbari, F. (2009). Prediction of project outcome: The application of statistical methods to earned value management and earned schedule performance indexes. *International Journal of Project Management, 27*(4), 400–407.

Meredith J. R., & Mantel S. J. (2015). *Project management—A managerial approach* (9th ed.). Wiley.

Mintzberg, H. (1994). The fall and rise of strategic planning. *Harvard Business Review, 72*(1), 107–114.

Musawir, A., Serra, C. E. M., Zwikael, O., & Ali, I. (2017). Project governance, benefit management, and project success: Towards a framework for supporting organizational strategy implementation. *International Journal of Project Management, 35*(8), 1658–1672.

OGC (Office of Government Commerce). (2009). Managing successful projects with PRINCE2. The Stationery Office: London, UK.

OGC. (2011). *Managing successful programmes*. London, UK: The Stationery Office.

Pinto, J. K., & Mantel, S. J. (1990). The causes of project failure. *IEEE Transactions on Engineering Management, 37*(4), 269–276.

Pinto, J. K., & Slevin, D. P. (1989). Critical success factors in R&D projects. *Research Technology Management, 32*(1), 31–35.

PMI Standards Committee. (2017). *A guide to the project management body of knowledge* (6th ed.). Newtown Square.

Rad, P. F., & Levin, G. (2002). The advanced project management office: A comprehensive look at function and implementation. Boca Raton, FL: St. Lucie Press.

Rubin, I. M., & Seeling, W. (1967). Experience as a factor in the selection and performance of project managers. *IEEE Transactions Engineering, 14*(3), 131–134.

Ryder-Smith, J. (1998). The secret of good conversation—Investing in success. *Health Manpower Management, 24*, 38–39.

Scott-Young, C., & Samson, D. (2008). Project success and project team management: Evidence from capital projects in the process industries. *Journal of Operations Management, 26*(6), 749–766.

Shenhar, A. J., & Dvir D. (2007). *Reinventing project management: The diamond approach to successful growth and innovation*. Harvard Business School Press.

Shenhar, A., & Holzmann, V. (2017). The three secrets of megaprojects success: Clear strategic vision, total alignment, and adapting to complexity. *Project Management Journal, 48*(6), 29–46.

Simon, H. (1976). *Administrative behavior* (3rd ed.), New York: The Free Press.

Smyrk, J. (1995). The ITO model: A framework for developing and classifying performance indicators. In *The international conference of the Australasian evaluation society*. Sydney, Australia.

Stapleton, T. (1993). On track with the metro line megaproject in Los Angeles. In *Business forum* (Vol. 18, No. 3, pp. 10–13). Los Angeles: California State University.

Symons, C. (2010). Software industry performance: What you measure is what you get. *IEEE Software, 27*(6), 66–72.

Tichy, L., & Bascom, T. (2008). The business end of IT project failure. *Mortgage Banking, 68*(6), 28–35.

Turner, J. R., & Muller, R. (2004). Communication and co-operation on projects between the project owner as principal and the project manager as agent. *European Management Journal, 22*(3), 327–336.

Unger-Aviram, E., Zwikael, O., & Restubog, S. (2013). Revisiting goals, feedback, recognition and performance success: The case of project teams. *Group & Organization Management: An International Journal, 38*(5), 570–600.

Walker, D. H. T., & Nogest, K. (2008). Performance measures and project procurement. In D. H. T. Walker & S. Rowlinson (Eds.), *Procurement systems—A cross industry project management perspective*. London: Taylor & Francis.

Ward, L. (2000). Project management terms—A working glossary (2nd ed.). ESI International.

Weick, M., & Guinote, A. (2010). How long will it take? Power biases time predictions. *Journal of Experimental Social Psychology 46*, 595–604.

Wright, J. K., Carpenter, E. W., Hunt, A. G., & Downhill, B. (1958). Observations on the explosion at ripple rock. *Nature, 182, 4649*, 1597–1598.

Zwikael, O. (2008). Top management involvement in project management—Exclusive support practices for different project scenarios. *International Journal of Managing Projects in Business, 1*(3), 387–403.

Zwikael, O. (2009). Critical planning process in construction projects. *Construction Innovation, 9* (4), 372–387.

Zwikael, O. (2009). The relative importance of the PMBOK® Guide's nine knowledge areas during project planning. *Project Management Journal, 40*(4), 94–103.

Zwikael, O., & Ahn, M. (2011). The effectiveness of risk management: An analysis of project risk planning across countries and industries. *Risk Analysis: An International Journal, 31*(1), 25–37.

Zwikael, O., Chih, Y., & Meredith, J. (2018). Project benefit management: Setting effective target benefits. *International Journal of Project Management, 36*, 650–658.

Zwikael, O., Elias, A., & Ahn, M. (2012). Stakeholder collaboration and engagement in virtual projects. *International Journal of Networking and Virtual Organisations, 10*(2), 117–136.

Zwikael, O., & Globerson, S. (2004). Evaluating the extent of use of project planning: A model and field results. *International Journal of Production Research, 42*(8), 1545–1556.

Zwikael, O., & Globerson, S. (2006). From critical success factors to critical success processes. *International Journal of Production Research, 44*(17), 3433–3449.

Zwikael, O., & Globerson, S. (2007). Quality management—A key process in the service industries. *Service Industries Journal, 27*(8), 1007–1020.

Zwikael, O., Globerson, S., & Raz, T. (2000). Evaluation of models for forecasting the final cost of a project. *Project Management Journal, 31*(1), 53–57.

Zwikael, O., Levin, G., & Rad, P. (2008). Top management support—The project friendly organization. *Cost Engineering Journal, 50*(9), 22–30

Zwikael, O., & Meredith, J. (2018). Who's who in the project zoo? The ten core project roles. *International Journal of Operations & Production Management, 38*(2), 474–492.

Zwikael, O., & Sadeh, A. (2007). Planning effort as an effective risk management tool. *Journal of Operations Management, 25*(4), 755–767.

Zwikael, O., Shimizu, K., & Globerson, S. (2005). Cultural differences in project management processes: A field study. *International Journal of Project Management, 23*(6), 454–462.

Zwikael, O., Shtub, A., & Chih, Y. (2015). Simulation-based training for project management education: Mind the gap, as one size does not fit all. *Journal of Management in Engineering, 31*(2), 040140351-11.

Zwikael, O., & Smyrk, J. (2012). A general framework for gauging the performance of initiatives to enhance organizational value. *British Journal of Management, 23*, S6–S22.

Zwikael, O., & Smyrk, J. (2015). Project governance: Balancing control and trust in dealing with risk. *International Journal of Project Management, 33*(4), 852–862.

Index

O. Zwikael and J. R. Smyrk, *Project Management*,
https://doi.org/10.1007/978-3-030-03174-9

The manufacturer's authorised representative in the EU is Springer Nature Customer Service Centre GmbH, Europaplatz 3, 69115 Heidelberg, Germany. If you have any concerns regarding our products, please contact ProductSafety@springernature.com

Printed and bound by CPI Group (UK) Ltd, Croydon, CR0 4YY
30/09/2026
02225444-0015